MODELING AND PRECISION CONTROL OF SYSTEMS WITH HYSTERESIS

MODELING AND PRECISION CONTROL OF SYSTEMS WITH HYSTERESIS

LEI LIU and
YI YANG

AMSTERDAM • BOSTON • HEIDELBERG • LONDON
NEW YORK • OXFORD • PARIS • SAN DIEGO
SAN FRANCISCO • SINGAPORE • SYDNEY • TOKYO
Butterworth-Heinemann is an imprint of Elsevier

Butterworth Heinemann is an imprint of Elsevier
The Boulevard, Langford Lane, Kidlington, Oxford OX5 1GB, UK
225 Wyman Street, Waltham, MA 02451, USA

Notices
Knowledge and best practice in this field are constantly changing. As new research and experience broaden our
understanding, changes in research methods, professional practices, or medical treatment may become necessary.

Practitioners and researchers must always rely on their own experience and knowledge in evaluating and
using any information, methods, compounds, or experiments described herein. In using such information or methods
they should be mindful of their own safety and the safety of others, including parties for whom they have a
professional responsibility.

To the fullest extent of the law, neither the Publisher nor the authors, contributors, or editors, assume any liability
for any injury and/or damage to persons or property as a matter of products liability, negligence or otherwise,
or from any use or operation of any methods, products, instructions, or ideas contained in the material herein.

British Library Cataloguing in Publication Data
A catalogue record for this book is available from the British Library

Library of Congress Cataloging-in-Publication Data
A catalog record for this book is available from the Library of Congress

ISBN: 978-0-12-803528-3

For information on all Butterworth Heinemann publications
visit our website at http://store.elsevier.com/

Working together
to grow libraries in
developing countries

www.elsevier.com • www.bookaid.org

Publisher: Joe Hayton
Acquisition Editor: Sonnini R. Yura
Editorial Project Manager: Mariana Kühl Leme
Editorial Project Manager Intern: Ana Claudia A. Garcia
Production Project Manager: Kiruthika Govindaraju
Marketing Manager: Louise Springthorpe
Cover Designer: Harris Greg

CONTENTS

PREFACE

Smart systems are increasingly applied in precision engineering applications because of their capability of accurate motion, such as in scanning-probe-microscopy-based nanofabrication and dynamic imaging of molecules, piezoelectric system-based ultra-precision, and high-stability pointing of aerospace optical sensitive instruments. Currently, dynamic motion tracking of smart systems is still not sufficient because of their hysteretic dynamics.

This book studies systems with hysteresis from an engineering perspective. The fundamentals of systems with hysteresis are described. Static hysteresis effect, mechanical vibrations dynamics, electric dynamics, and creep effect are investigated. Then, hysteresis modeling in smart actuators is presented. Comprehensive modeling of multifield hysteretic dynamics in smart systems involving the synthesis of hysteresis and nonhysteretic dynamics is also presented. Next, control approaches, such as proportional-integral-derivative tuning control, inversion-based feedforward control, robust control, composite control, and multirate control, are investigated for smart systems. Finally, two cases studies of smart systems are presented. The modeling and precision control are applied in the case of a fast-steering mirror and an active vibration isolator.

This book offers engineers fundamental modeling and precision control approaches for smart systems with hysteresis. The book is primarily concerned with the applications of modeling and precision control in various situations.

Northwestern Polytechnical University,
China
Associate Professor
Lei Liu

Luoyang Institute of Electro-optical Devices,
China
Dr Yi Yang

ACKNOWLEDGMENTS

We express our sincerest gratitude and appreciation to all who helped us prepare this book.

This book is supported by the Enterprise-University-Research Cooperation Innovation Project of the Aviation Industry Corporation of China and the National Natural Science Foundation of China (11402044).

We especially thank our families for their continued support.

INTRODUCTION

In this chapter, we review the modeling and control of smart systems with hysteresis. First, the background and applications are introduced. Next, the modeling, identification, and control algorithms of systems with hysteresis are reviewed. Then, some key problems in systems with hysteresis are discussed. Finally, the content of this book is summarized.

In this book we investigate smart systems with hysteresis. The smart systems are typically driven by smart actuators, such as piezoelectric actuators, shape-memory-alloy actuators, or magnetostrictive actuators. Compared with conventional motors, smart actuators could provide more precise motion.

1.1   Motivation

Smart systems are increasingly applied in precision engineering. For example, scanning probe microscopes, atomic force microscopes, and positioning XYZ stages can provide nanometer-accuracy positioning and kilohertz scanning. In aerospace applications, piezoelectric smart systems are increasingly demanded in active vibration isolators and active optics (resolution enhancement and imaging stabilization), such as for a Stewart vibration isolator, a fast-steering mirror (line-of-sight jitter control), or a deformable mirror (wavefront control).

Currently, precision and high-bandwidth motion are required simultaneously. Classic feedback is not able to satisfy the strict specifications. Accurate modeling and control approaches are demanded by various engineers. Thus, in this book we present modeling and precision control of smart systems with hysteresis. We try to provide engineers with fundamental knowledge.

1.2 Literature Review

Piezoelectric and other smart materials are increasingly used in precision engineering, such as in scanning probe microscopes in the life sciences and nanomanufacturing, precision active optics in astronomy, space laser communication, and space imaging cameras, and microelectromechanical systems, but the dynamic performance is limited by the hysteretic dynamics, which involves to multiple fields and multiple timescales. For example, the operating bandwidth of scanning probe microscopes with fine tracking is less than 1-5% of the first resonant frequencies.

To overcome the bandwidth limitation and provide both precision and high-speed motion, accurate modeling and control are required for smart systems with hysteresis. Various modeling approaches regarding hysteresis theory have been investigated. Some remarkable monographs are *Systems with Hysteresis* by Krasnosel'skii and Pokrovskii [1], *Mathematical Models of Hysteresis and Their Applications* by Mayergozy [2], and *Differential Models of Hysteresis* by Visintin [3]. In *Systems with Hysteresis*, Krasnosel'skii and Pokrovskii present systematic mathematical modeling of systems with hysteresis, especially the hysteron theory, and the identification, control, and stability problem. In *Mathematical Models of Hysteresis and Their Applications*, Mayergozy presents detailed mathematical models of hysteresis nonlinearities, especially the Preisach model, as well as the hysteresis properties and physical meaning. In *Differential Models of Hysteresis*, Visintin presents various methods for analyzing hysteresis models, such as the Prandtl, Ishlinskii, Preisach, and Duhem models. Partial differential equations are also introduced together with hysteresis operators. These monographs greatly improve the level of theory of systems with hysteresis, but they may still be difficult to understand for engineers in precision engineering.

The control approaches for smart systems are very colorful. Various control techniques have been investigated for smart systems, such as classic control, robust control, inversion-based feedforward control, repetitive control, adaptive control, intelligent control, iterative learning control, and neural network-based control. More details of the literature review are contained in each chapter. Currently, the modeling and control approaches for smart systems with hysteresis may be very complex. It may be difficult for engineers to understand the hysteresis theory. Thus, in this book we propose modeling and precision control from an engineering perspective.

1.3 Book Objectives

Smart materials exhibit hysteretic dynamics. To overcome the hysteresis over a broad range of frequencies, it is necessary to investigate the precision modeling and control of systems with hysteresis, which is greatly demanded by smart systems developers. Current books on systems with hysteresis are not sufficient for engineers to develop smart systems. Precision modeling and control of systems with hysteresis are increasingly demanded by engineers. Thus, in this book we investigate the modeling and control of systems with hysteresis from an engineering perspective.

1.4 Book Overview

This book is organized as follows. In Chapter 2, the fundamentals of systems with hysteresis are discussed. In Chapter 3, the hysteresis model in smart actuators is presented. In Chapter 4, a comprehensive model of multifield hysteretic dynamics is presented. In Chapter 5, control approaches for systems with hysteresis are investigated. In Chapter 6, a case study of a piezoelectric steering platform is presented. Finally, in Chapter 7, a case study of active vibration isolation is presented.

References

[1] M.A. Krasnosel'skii, A.V. Pokrovskii, Systems with Hysteresis, Springer-Verlag, New York, 1989 (translated from Russian by M. Niezgodka).
[2] I.D. Mayergozy, Mathematical Modeling of Hysteresis, Springer-Verlag, New York, 1991.
[3] A. Visintin, Differential Models of Hysteresis, Springer-Verlag, New York, 1994.

2

FUNDAMENTALS OF SYSTEMS WITH HYSTERESIS

Abstract

In this chapter we introduce the fundamental knowledge of systems with hysteresis. The main dynamics and effects are discussed, such as the static (rate-independent) hysteresis effect, the dynamic hysteresis effect, the creep effect, mechanical vibration, electrical dynamics, and the composite model. The properties of systems with hysteresis are presented in detail. Knowledge of the proposed dynamics, effects, and their properties is necessary to investigate systems with hysteresis. Especially, we propose modeling and control approaches for piezoelectric systems. To make the modeling easier to understand, we provide various figures and experimental data.

Keywords: Smart systems, Hysteresis, Preisach model, Rate independence, Dynamic hysteresis

2.1 Introduction

Smart systems are increasingly demanded in various fields where ultraprecision motion and steering play a role. More and more engineers from different backgrounds (such as electrical, mechanical, optical, and automatic control, life sciences, and precision instruments) are beginning to design smart systems. It is important to provide fundamental modeling and control approaches for various engineers. The purpose of this chapter is to present the fundamentals of systems with hysteresis, and to illustrate how to represent smart systems with suitable methods.

Systems with hysteresis are very complex. Theoretically, the hysteresis has global memory, which is different from typical ordinary differential equations.

2.2 Smart Systems with Hysteresis

Smart systems exhibit hysteresis due to smart materials such as piezoelectric material and shape-memory-alloy material. Four types of models are commonly used to represent smart systems. First, mechanical vibration dynamics are widely used to represent smart systems, as shown in Figure 2.1. In this model, the hysteresis, creep, and electrical dynamics are not considered, but the modeling accuracy could be acceptable to regulate basic controllers.

In some nonaccurate situations, capacitor dynamics can also be used to represent smart systems, as shown in Figure 2.2. For typical systems driven by smart materials, the operating voltage

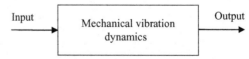

Figure 2.1 Mechanical vibration dynamics.

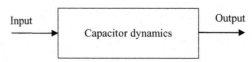

Figure 2.2 Capacitor dynamics.

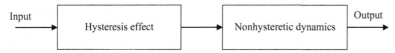

Figure 2.3 Hysteretic dynamics with a series connection.

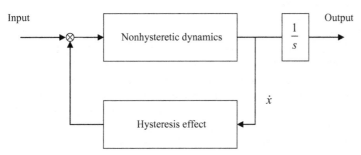

Figure 2.4 Hysteretic dynamics with a feedback connection.

can be as high as 1 kV, but the current is as small as 1 mA. The direct current is approximately zero, and the whole system behaves as a capacitor driven by a voltage source. For all commercial piezoelectric actuators, the capacitive value should be specified. Modeling examples of the mechanical and capacitor dynamics will be presented in Section 2.3.

In the former two representations, the hysteresis effect is not modeled, and this may be acceptable in modest smart systems. In most cases, high accuracy is achieved by reduction of speed and bandwidth. Recently, high speed and high accuracy have been required simultaneously in smart systems. In these situations, accurate modeling and control approaches are demanded. To simplify the model, the hysteretic dynamics by series or feedback combinations are increasingly used to represent smart systems as shown in Figures 2.3 and 2.4. It is easier to understand and manipulate the series connection, as shown in Figure 2.3. The hysteresis effect can be represented separately by the Preisach model, the Prandtl-Ishlinskii model, the Maxwell model, etc. The nonhysteretic dynamics comprise mechanical vibrations, electrical dynamics, creep effects, etc. The hysteresis effect and the nonhysteretic dynamics are weakly coupled. The identification and compensation of the series connection is also easier to implement, because the hysteresis effect and the nonhysteretic dynamics can be identified and compensated step-by-step.

Figure 2.4 shows the model of smart systems with hysteresis obtained with a feedback connection. Generally, the velocity is

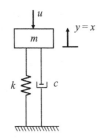

Figure 2.5 Mechanical dynamics.

used for feedback. $1/s$ represents an integral action. The velocity of the smart systems is transferred to the displacement. In the feedback connection, the whole model strut is compact, and the mathematics is stricter. However, identification and compensation are more difficult because the couplings in feedback hysteresis systems are more serious.

2.3   Mechanical and Capacitor Dynamics

Figure 2.5 shows the mechanical dynamics of smart systems: u represents the input force, which is derived from the input voltage; x denotes the output displacement, and $y = x$; k and c represent the stiffness and damping of the smart system, respectively.

The transfer function from the input force to the output displacement can be written as

$$\frac{Y(s)}{U(s)} = \frac{1}{ms^2 + cs + k}, \tag{2.1}$$

where m, c, and k represent the equivalent mass, damping, and stiffness of the smart system, respectively.

Moreover, if we let ξ_n and ω_n denote the damping ratio and the natural frequency, respectively, then (2.1) can be rewritten as

$$\frac{Y(s)}{U(s)} = \frac{1/m}{s^2 + 2\xi_n\omega_n s + \omega_n^2}. \tag{2.2}$$

Further, more mechanical vibration modes could be modeled as in the following equation:

$$\frac{Y(s)}{U(s)} = k\sum_i = 1^N \frac{\omega_i^2}{s^2 + 2\xi_i\omega_i s + \omega_i^2}, \tag{2.3}$$

where N denotes the order of the vibrational modes, and ξ_i and ω_i represent the damping ratio and the mode frequency of vibrational mode i.

Figure 2.6 shows the capacitor dynamics, in which u and y represent the input voltage and output displacement, C represents the capacitance of the piezoelectric actuator, PA denotes the piezoelectric actuator, and R represents the resistance due to

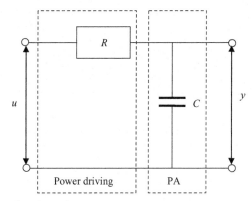

Figure 2.6 Capacitor dynamics.

power driving. Equation (2.4) denotes the transfer function from the input voltage u to the output displacement y:

$$\frac{Y(s)}{U(s)} = \frac{1}{RCs + 1}, \tag{2.4}$$

where RC denotes the time constant of the smart system. This parameter indicates the capacitor behaviors of smart systems.

Conveniently, RC is rewritten as τ. Then, the capacitor dynamics can be rewritten as

$$\frac{Y(s)}{U(s)} = \frac{1}{\tau s + 1}. \tag{2.5}$$

2.4 Static Hysteresis

At low frequencies, hysteresis is the main factor degrading the precision of smart actuators, and the Preisach model can be used to describe this non-local memory nonlinearity. The double integral and memory characteristic of this nonlinearity is represented geometrically in the Preisach plane, and illustrates that only the past input extreme values and the current input will determined the current and future hysteresis output. The wiping-out property, the rate-independence property, and congruence property of the Preisach model of hysteresis are analyzed and simulated. To allow practical identification of this model, the Preisach plane is represented as a collation of discrete lattices with different weights/densities. With such a discretized representation, parameter identification can be done efficiently, and simulation shows that the relative error of the model's output with respect to the real output is less than 5%.

Among nonlinearities present in piezoelectric systems, hysteresis contributes the main uncertainty which affects the control or measurement performance. In the open loop, the maximum error from hysteresis is 10-15% of the total displacement of the piezoelectric actuator [1].

Hysteresis models can be classified into physical models [2–4] and mathematical models. Physical models are constructed on the basis of physical laws applied to the phenomenon of hysteresis, and thus models are intuitive but may be in a complex form which is difficult to identify. Conversely, mathematical models are a vehicle to provide a relationship among variables close to the actual system and they are usually more amenable to practical use for identification and control. In the current literature, the ferromagnetic hysteresis model [2] is a physical model and the Prandtl-Ishlinskii model [5, 6] and the Preisach model are two popular mathematical models [2]. The Prandtl-Ishlinskii model has been used to compensate the hysteresis nonlinearity of piezoelectric actuators [7], the Bouc-Wen model has been applied to describe the nonlinearity of piezoelectric actuators [8], and the distribution weight function has been assumed to be of a specific form to reduce the number of measurements [7].

2.4.1 Preisach Hysteresis

The hysteresis relay, which constitutes the basic element of hysteresis in piezoelectric (PZT) actuators is as shown in Figure 2.7 where the output $\gamma_{\alpha,\beta}$ of the operator is assumed to be of two values (0 and 1), and α and β are the switching threshold values of the hysteresis operator $\gamma_{\alpha,\beta}[u(t)]$. The ascending branch is traced when the input is monotonically increasing, and the descending branch is traced when the input is monotonically decreasing. The hysteresis output (i.e., the voltage-to-displacement dynamics) is given in (2.6):

$$f(t) = \iint_{\alpha>\beta} r_{\alpha\beta}[u(t)] \, \mathrm{d}\alpha \, \mathrm{d}\beta, \qquad (2.6)$$

where $\mu(\alpha, \beta)$ is the distribution weight function and $f(t)$ is the piezoceramic expansion. In the Preisach model, the same input $u(t)$ is first applied to all the hysteresis operators $\gamma_{\alpha,\beta}[u(t)]$. The output of these operators is weighted by a Preisach function $\mu(\alpha, \beta)$ unique to the actuator, and then summed continuously over possible values of α and β, as shown in Figure 2.7. The

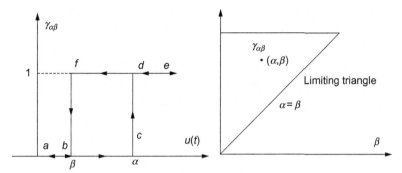

Figure 2.7 Hysteresis operator and the Preisach plane.

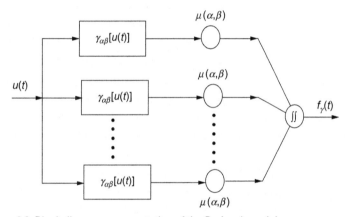

Figure 2.8 Block diagram representation of the Preisach model.

hysteresis output can thus be considered to be a superposition of a continuous set of two-position relay operators $\gamma_{\alpha,\beta}[u(t)]$ over the range of input signal. The physical meanings of Preisach model in (2.6) are illustrated in Figure 2.8.

2.4.2 Preisach Plane

The Preisach plane is shown on the right in Figure 2.7. The double integration in (2.6) is performed over the range $u_{\min} \leq \beta \leq \alpha \leq u_{\max}$, where $u_{\min}$ and $u_{\max}$ represent the minimum and maximum values of the input to the PZT actuator, respectively. The maximum and minimum values of the input and the switching condition $\alpha \geq \beta$ give rise geometrically to a limiting triangle on the Preisach plane where each pair of (α, β) in the plane defines a unique operator $\gamma_{\alpha,\beta}[u(t)]$. The limiting triangle corresponds to

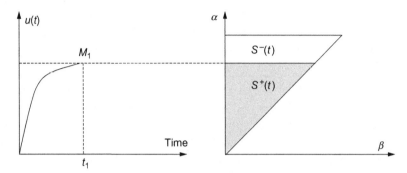

Figure 2.9 Input signal and the limiting triangle in the Preisach plane.

the major loop of hysteresis, and the Preisach function $\mu(\alpha, \beta)$ is defined to be zero outside the limiting triangle.

Each movement in the input signal will shape the limiting region in the Preisach plane accordingly. Figure 2.9 illustrates this equivalence. As the input is monotonically increased to a value M_1, all the hysteresis operators $\gamma_{\alpha,\beta}[u(t)]$ with switching values less than M_1 will be activated. Geometrically, this corresponds to a division of the limiting triangle into two regions: S^+, wherein the hysteresis operators are activated, and S^{-1}, wherein the hysteresis operators are deactivated. The interface link between S^+ and S^{-1} is a horizontal line given by the line equation, as shown in Figure 2.9.

Next, when the input starts to monotonically decrease from M_1 to m_1, all the activated hysteresis operators earlier with switching values larger than m_1 will deactivate. This means that the previous subdivision of the limiting triangle is further changed, and the interface link moves from right to left, with the coordinates of its vertex being (M_1, m_1). Consequently, at this time, a smaller limiting region has evolved corresponding to the shaded region in Figure 2.10.

The interface link between the two regions possesses vertex coordinates that correspond to the local maxima and local minima of the input. From this way of interpreting the input-output relationship, we can rewrite the classical Preisach model as

$$f(t) = \iint\limits_{S^+(t)} \mu(\alpha, \beta) \, d\alpha \, d\beta, \tag{2.7}$$

which is a weighted computation of the area of the limiting region thus evolved.

After some time, the input monotonically increases from m_1 to $M_2 (M_2 < M_1)$, as shown in Figure 2.11.

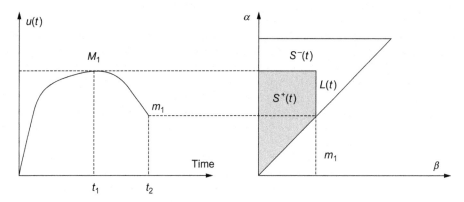

Figure 2.10 Input signal and limiting area representation in the Preisach plane.

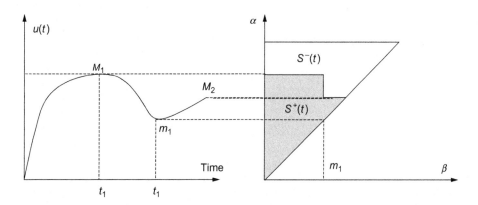

Figure 2.11 Input signal and limiting area representation in the Preisach plane.

Finally, to illustrate the input-output relationship under this model, consider an input signal as shown in Figure 2.12 (left) with model parameters $\alpha_{max} = 10, \alpha_{min} = 0, \beta_{max} = 10, \beta_{min} = 0$, and $\mu = 4$. The corresponding hysteresis output is shown on the right in Figure 2.12 (right).

2.4.3 Preisach Hysteresis Properties

The Preisach model is useful for its ability to replicate characteristics of hysteresis observed in PZT actuators. In the following sections, these properties are highlighted from the equivalent limiting region evolution in the Preisach plane.

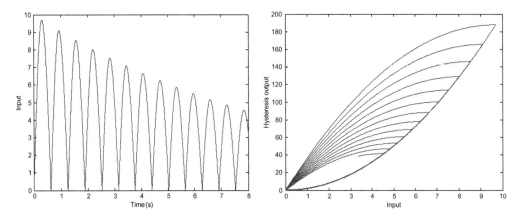

Figure 2.12 Input and hysteresis loop of the hysteresis model.

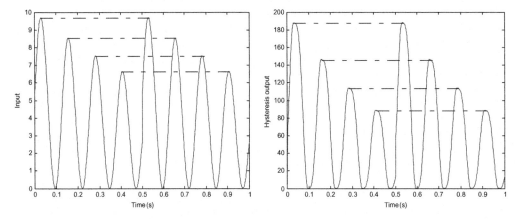

Figure 2.13 Input voltage $u(t)$ and output (no units) of the Preisch hysteresis.

Wiping-out Property

In the Preisach plane, each local maximum value of input $u(t)$ will wipe out the vertices of the limiting boundary $L(t)$ whose α values are less than $u(t)$, and each local minimum value of input $u(t)$ wipes out the vertices of the limiting boundary $L(t)$ whose β values are more than $u(t)$. The wiping-out properties of hysteresis mean that only dominant extreme values are stored in the Preisach model and all the other input local extreme values are wiped out. The wiping-out property is shown in Figures 2.13 and 2.14, in which the global extreme value emerges at 0.53 s, such that all the input extreme values before $t = 0.53$ s which are no more than the extreme value at $t = 0.53$s are wiped out in the Preisach plane

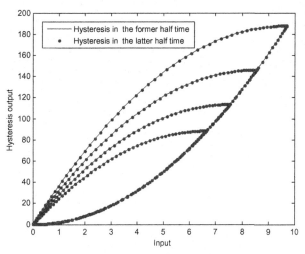

Figure 2.14 Input voltage $u(t)$ and hysteresis loops of the Preisch hysteresis.

and they do not affect the future output. In Figure 2.13, it can be seen that the hysteresis output after $t = 0.53$ s is independent of the input before $t = 0.53$ s. In other words, the outputs in the time intervals [0.3, 0.53] and [0.53, 1.3] with the same input signal series share the same output signal series because of the same global maximum values of the input signal at $t = 0.53$ s. The property is also shown in Figure 2.14.

Rate-Independence Property

Static hysteresis models such as the classical Preisach model and the Prandtl-Ishlinskii model have the property of rate independence with respect to the input voltage at certain frequency bands, which means that the hysteresis branches are determined purely by the input extreme values, and the input rates do not affect the hysteresis loop. As shown in Figure 2.15, the rate of the input signal increases by 4 times at $t = 0.5$ s. However, the hysteresis loops holds. It indicates that the hysteresis loops of the Preisach model hold only if the input frequency is changed—that is, the static Preisach model is rate independent.

Congruence Property

Inputs oscillating identically between two extreme values give rise to congruent minor hysteresis loops. Two input signal series are used to generate the hysteresis output in this illustration. They have a different history in the time interval [0, 0.5]. However, they

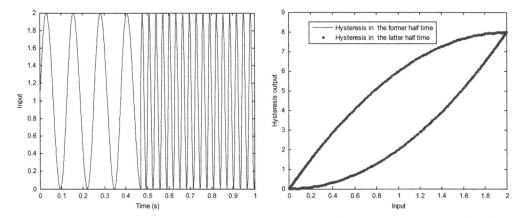

Figure 2.15 Rate-independence property.

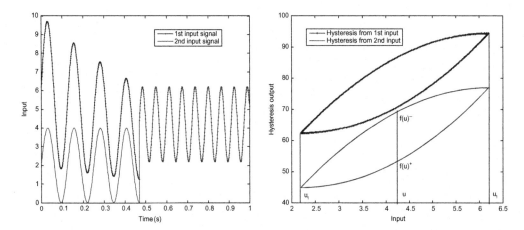

Figure 2.16 Congruent minor loops.

converge to the same form after $t = 0.5\,\text{s}$. The hysteresis minor loops obtained after $t = 0.5\,\text{s}$ as shown in Figure 2.16 have the same shape but are offset vertically (i.e., they are congruent).

We can illustrate this property by considering the limiting region after $t = 0.5\,\text{s}$. First, the input is set to u, and the input oscillations are assumed to lie between u_1 and u_2. The hysteresis output is $f(u)^+$ when the input voltage is increasing monotonously from u_1 to u. Conversely, the hysteresis output is $f(u)^-$ when the input is decreasing monotonously from u_2 to u. As shown in Figure 2.16, the variance of the hysteresis output when the input varies from u_1 to u is written as

$$f(u)^+ - f(u_1) = \iint\limits_{T_1} \mu\,(\alpha, \beta)\,\mathrm{d}\alpha\,\mathrm{d}\beta, \tag{2.8}$$

where $f(u_1)$ is the hysteresis output when the input is u_1, and T_1 is the limiting triangle with $u_1 \leqslant \alpha \leqslant u$, $u_1 \leqslant \beta \leqslant u$.

The variance of the hysteresis output following input path $u_1 - u - u_2 - u$ is rewritten as

$$f(u)^- - f(u_1) = \iint\limits_{T_2} \mu\,(\alpha, \beta)\,\mathrm{d}\alpha\,\mathrm{d}\beta, \tag{2.9}$$

where T_2 is the limiting trapezoid with four vertices (u_1, u), (u, u), (u_1, u_2), (u, u_2). Substituting (2.8) into (2.9), we obtain the hysteresis output change with input path $u - u_2 - u$, which is

$$
\begin{aligned}
f(u)^- &- f(u)^+ \\
&= \iint\limits_{T_2} \mu\,(\alpha, \beta)\,\mathrm{d}\alpha\,\mathrm{d}\beta - \iint\limits_{T_1} \mu\,(\alpha, \beta)\,\mathrm{d}\alpha\,\mathrm{d}\beta \\
&= \left(\iint\limits_{T} \mu\,(\alpha, \beta)\,\mathrm{d}\alpha\,\mathrm{d}\beta + \iint\limits_{T_1} \mu\,(\alpha, \beta)\,\mathrm{d}\alpha\,\mathrm{d}\beta \right) - \iint\limits_{T_1} \mu\,(\alpha, \beta)\,\mathrm{d}\alpha\,\mathrm{d}\beta \\
&= \iint\limits_{T} \mu\,(\alpha, \beta)\,\mathrm{d}\alpha\,\mathrm{d}\beta,
\end{aligned}
\tag{2.10}
$$

where T is the limiting rectangle with $u_1 \leqslant \alpha \leqslant u$, $u \leqslant \beta \leqslant u_2$.

Note that the vertical variance depends only on u, u_1, and u_2 and it is independent of the input history before the time corresponding to the convergence to the same form. This means that the shape of the minor loop between u_1 and u_2 is fixed, independently of the input history before the time at when $u = u_1$, and the input rate between u_1 and u_2.

2.5 Behavior Comparison of Preisach Hysteresis and Phase Delay under Sinusoidal Inputs

In this section, the behavior of Preisach hysteresis and that of phase delay under sinusoidal inputs are compared. The phase delay results from linear dynamics with a damping ratio of 0.2, a resonant frequency of 50 Hz, and a direct current gain of 1.

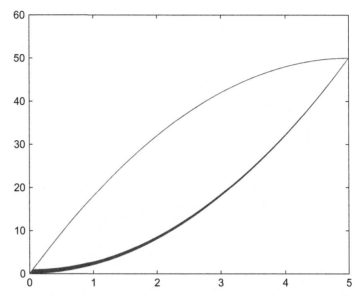

Figure 2.17 The hysteresis from the Preisach hysteresis model.

Figure 2.17 shows the input-output loop of Preisach hysteresis. Figures 2.19 and 2.20 show the input-output loops of the linear dynamics at 20 and 50 Hz, respectively. It can be seen that it is difficult to distinguish the input-output loops of the Preisach hysteresis and the linear dynamics: they have a similar input-output curve for a sinusoidal wave.

Further, Figure 2.18 shows the Bode diagram of the proposed linear dynamics. It can be seen that the phase delay of the linear system increases as the frequency increases from 0 to the natural frequency. The increasing phase delay can also be seen in Figures 2.19 and 2.20. After the natural frequency, the direction of the axis of hysteresis will change to 90°, resulting in one circular hysteresis loop. The observed phase delay of the linear dynamics makes the input-output loops more complex. As a result, it is difficult to investigate the linear dynamics from the point of Preisach hysteresis. In the diagrams of the input-output loops, the linear dynamics can behave similarly as a dynamic hysteresis. In experiments on the dynamic hysteresis proposed in various articles, it is shown that the hysteresis will become serious as the input frequency increases. Some composite representations that use Preisach hysteresis and linear dynamics are demanded to denote the complex dynamic hysteresis.

Figure 2.21 shows the input-output loop of linear dynamics at 100 Hz (twice the resonant frequency). It can be seen that the input-output loop varies significantly as the input frequency

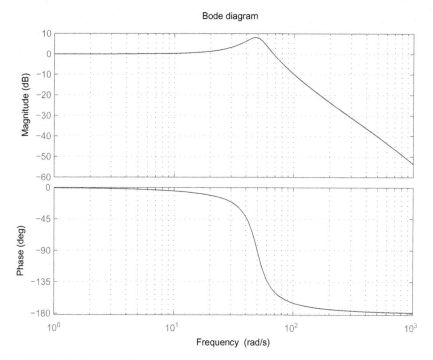

Figure 2.18 Bode diagram of linear dynamics.

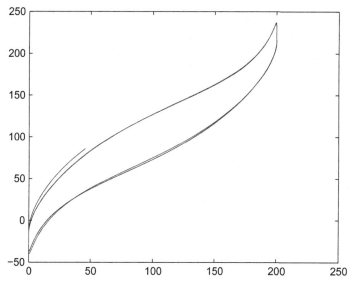

Figure 2.19 Input-output loop from second-order linear dynamics at the natural frequency.

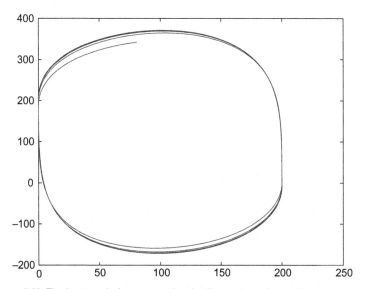

Figure 2.20 The hysteresis from second-order linear dynamics at the resonant frequency (50 Hz).

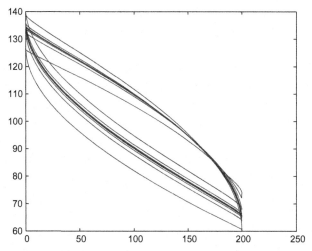

Figure 2.21 Input-output loop of the second-order linear dynamics at 100 Hz (twice the resonant frequency).

increases, but the input-output loop of the Preisach hysteresis holds as the input frequency increases—that is, Preisach hysteresis is rate independent.

The lag of rate-independent static hysteresis is different from that in linear dynamics. Figure 2.22 illustrates the lag in the rate-independent hysteresis and linear dynamics. For Preisach

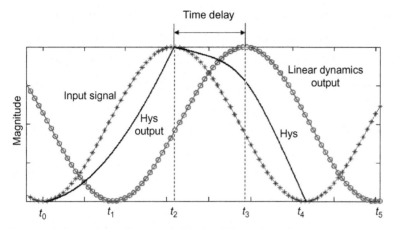

Figure 2.22 Comparison of hysteresis (Hys) and linear dynamics delay.

hysteresis, the output simultaneously approaches its maximum value when the input signal approaches its maximum value. This is different from the delay in linear dynamics, where the time of the maximum input is not equal to the time of the maximum output.

The effect of Preisach hysteresis under a square-wave input is to alter the magnitude of the square-wave input. In other words, under square-wave inputs, the classical Preisach hysteresis behaves like a nonlinear amplifier without a phase delay and dynamic response, as shown in Figure 2.23. Further, Figure 2.24 shows the input-output loop under a square-wave input. It can be seen that there is no hysteresis or delay.

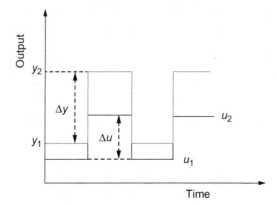

Figure 2.23 Static hysteresis response under a square-wave input.

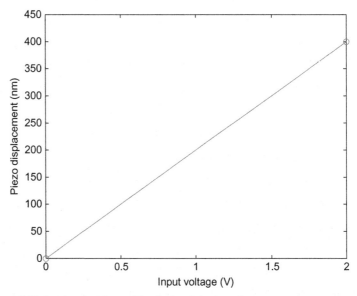

Figure 2.24 Input-output loop of the Preisach hysteresis.

2.6 Closed-Loop Response of Smart Systems with Hysteresis

In this section, the closed-loop responses of smart systems with hysteresis are presented. The results are collected with a piezoelectric XY stage by use of proportional-integral (PI) control. Capacitors are used to measure the microdisplacement on the order of nanometers.

In experiments, the PI controller is defined as in (2.11):

$$K_{PI} = k_p \left(1 + \frac{2.5}{\tau_i s}\right);$$ (2.11)

K_{PI} represents the PI controller, k_p is the gain of the PI controller, τ_i is the time constant of the PI controller. $k_P = 1.6 \times 10^5$, $\tau_i = 0.1$.

The set-point value is zero. Two initial values are used to test the closed-loop response. One initial value is 280 nm, and the other initial value is -4750 nm. The set-point tracking errors and control voltages of the two initial values in the experiment are shown in Figures 2.25 and 2.26, respectively.

Figure 2.25 shows the position response of the piezoelectric actuator system with different initial values. Figure 2.26 shows the control voltage response and the steady control voltage for the two cases are 1.4 and -0.2 V, respectively. The experimental results indicate that the steady output displacements are the same,

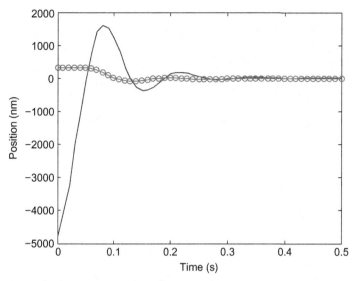

Figure 2.25 Displacement responses (the initial value for the solid line is 250 nm and the initial value for the line with circles is −4750 nm).

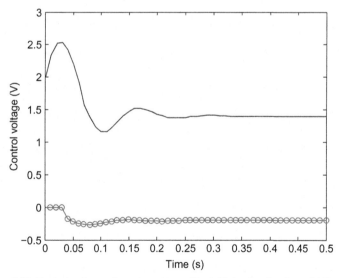

Figure 2.26 Control voltages in a closed loop (the initial value for the solid line is 250 nm and the initial value for the line with circles is −4750 nm).

but the steady control voltages with the same controller and set-point value are nonunique, which is a characteristic of hysteretic piezoelectric systems or other smart systems. The initial values of the smart systems result in different final control voltages even

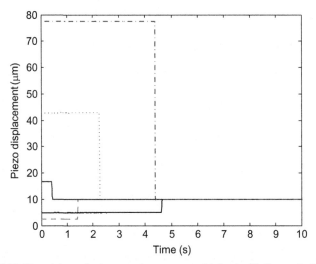

Figure 2.27 Closed-loop displacement responses obtained with the controller represented by (2.11) but with different initial values.

though the output displacements are the same. This characteristics is different from that of common systems represented by ordinary differential equations. In general, in nonmemory systems where disturbances are not considered, the final control voltages are the same if the set-point value and controllers are the same.

Furthermore, five different initial values have been tested with the PI controller represented by (2.11). The displacement responses of the five initial values are shown in Figure 2.27. Although the five initial values are different, the final steady displacement are the same with the PI controller represented by (2.11). Figure 2.28 shows the control voltages for the five initial values. Compared with Figure 2.27, the final steady control voltages for the five different initial values are different. This phenomenon demonstrates the existence of hysteresis in the smart systems.

2.7 Dynamic Hysteresis

Smart systems behave with static hysteresis at low frequencies, but over a broad range of frequencies, smart systems exhibit dynamic hysteresis. Experimental investigations verified that at broadband frequencies the voltage-to-displacement curve is strongly affected by the frequency of the input signal, and this kind of hysteresis is rate dependent. Figure 2.29 illustrates the dynamic hysteresis loops of the piezoelectric stage in experiments. We first measured the hysteresis loops at low range of frequencies (0.005-

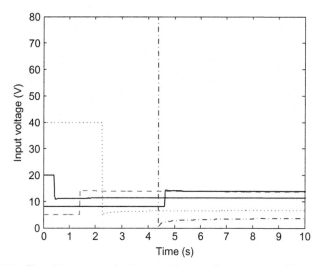

Figure 2.28 Closed-loop control voltages with controller represented by (2.11) but with different initial values.

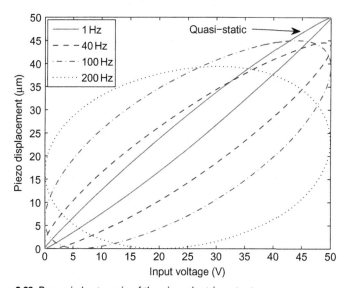

Figure 2.29 Dynamic hysteresis of the piezoelectric actuator.

0.1 Hz), and found the normalized hysteresis loops matched each other at different low frequencies (0.005-1 Hz); this is a typical rate-independent property. As the input frequency increases further, the hysteresis loops change shape, and even direction; this is thus a dynamics effect. Smart systems not only exhibit a hysteresis effect at low frequencies, but also exhibit a dynamic hysteresis effect over a broad range of frequencies.

Smart systems exhibit complex dynamic hysteresis. The quasi-static hysteresis model should be replaced with a dynamic type. One convenient approach to represent the dynamic hysteresis is to modify static hysteresis models. Mayergoyz [9] presented a dynamic Preisach model to describe the broadband hysteresis, but the parameter identification may be difficult for engineers. Janaideh [10] expanded the rate-independent Prandtl-Ishlinskii hysteresis to the rate-dependent type, and density functions were assumed for parameter identification. Instead of expanding the rate-independent hysteresis, one can use the cascade connection of the rate-independent hysteresis and the nonhysteretic dynamics to represent dynamic hysteresis over a broad range of frequencies [11].

Alternatively, we will use one composite representation in which the static hysteresis, mechanical vibrations, and electrical dynamics are connected in series such that the response of smart systems can be predicted over a broad range of frequencies. The composite representations of dynamic hysteresis will be presented in the next section.

Moreover, we test the open-loop responses using a stair-increasing input voltage and a stair-decreasing input voltage, as shown in Figures 2.30 and 2.31, respectively. It can be seen that smart systems show different responses under stair input

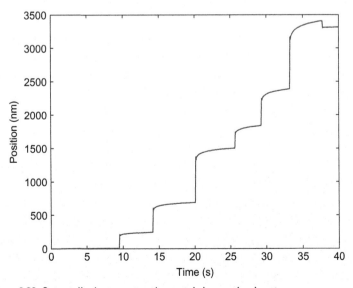

Figure 2.30 Output displacement under a stair-increasing input.

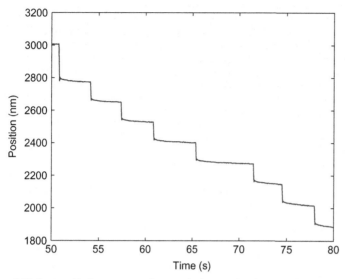

Figure 2.31 Output displacement under a stair-decreasing input.

voltages. In Figure 2.30, the mechanical vibration dynamics and electrical dynamics have faster responses, but the steady values keep increasing slowly after each step change. In Figure 2.31, the mechanical vibration dynamics and electrical dynamics also have faster responses, but the steady values keep dropping slowly after each step change. The responses in Figures 2.30 and 2.31 verify the complex dynamic hysteresis in smart systems.

2.8 Composite Representation of Dynamic Hysteresis

To represent the dynamic hysteresis with a convenient model, we use one composite representation based on static hysteresis, mechanical vibration, electrical dynamics, and the creep effect in series. The proposed composite representation is shown in Figure 2.32, in which the dynamic hysteresis can be represented at broadband frequencies, for broad amplitudes, and for multiple timescales. The mechanical vibration and electrical dynamics exhibit fast responses. For smart systems with high stiffness, the typical responses of the mechanical vibration and electrical dynamics are on the order of milliseconds. For smart systems with low stiffness, the typical responses of the mechanical vibration and electrical dynamics are on the order of 100 ms. The typical response of the creep effect in smart systems is on the order of

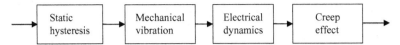

Figure 2.32 Composite representation of dynamic hysteresis.

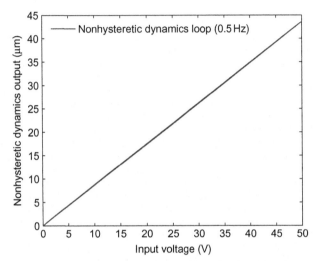

Figure 2.33 Nonhysteretic dynamics loop at 0.5 Hz in the piezoelectric actuator.

minutes. Models of static hysteresis, such as the Preisach model and Prandtl-Ishlinskii model, have no direct time variable. The static hysteresis is rate independent. The static hysteresis can respond as fast as the input voltage. Common dynamics behaves as a low-pass filter, but the static hysteresis behaves as a full-pass filter.

For a typical piezoelectric smart system with high stiffness, the responses of the mechanical vibration and electrical dynamics are on the order of milliseconds. Figure 2.33 shows the hysteresis loop at 0.5 Hz resulting from the mechanical vibration and electrical dynamics. It can be seen that the hysteresis loop is very narrow. The delay is also very weak. The response of the mechanical and electrical dynamics in a hard piezoelectric smart system is very weak.

Figure 2.34 shows the measured input-output (hysteresis) loop for a typical hard piezoelectric smart system. If the static hysteresis is not modeled, the hysteresis loop will contribute an input uncertainty to the mechanical and electrical dynamics. Figures 2.33 and 2.34 indicate that in a typical piezoelectric smart system, the responses at low frequency are mainly contributed by the static hysteresis, and the responses at high frequencies are mainly contributed by the mechanical and electrical dynamics.

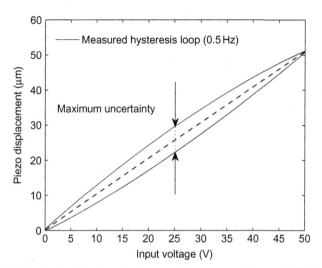

Figure 2.34 Nonhysteretic dynamics loop and the measured loop at 0.5 Hz in the piezoelectric actuator.

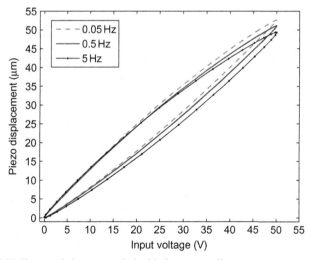

Figure 2.35 Hysteresis loops coupled with the creep effect.

Moreover, Figure 2.35 shows the hysteresis loops at a low range of frequencies. As the input frequency increases slowly, the hysteresis loop also drifts slowly. This indicates that the static hysteresis is coupled with the creep effect.

Further, Figure 2.36 shows typical responses of a hard piezo-electric smart system under square input voltages at a frequency of 10 Hz. The responses of the mechanical vibration, electrical dynamics, and creep effect, can be seen.

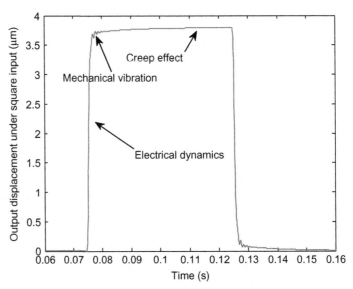

Figure 2.36 Output displacement of piezoelectric actuators under square inputs.

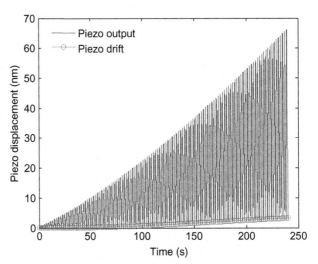

Figure 2.37 Piezoelectric displacement under sinusoidal inputs with increasing amplitude.

Figure 2.37 shows the responses of the piezoelectric system under a sinusoidal input with increasing amplitude. At the beginning the input voltage is zero. It can be seen that the output displacement slowly drifts owing to the creep effect. The circles corresponds to zero input voltage. According to the Preisach plane

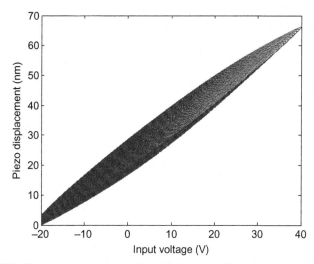

Figure 2.38 Hysteresis loops under sinusoidal inputs with increasing amplitude.

of the static hysteresis, the output displacement will return to zero after one sinusoidal period. The circle bias with respect to zero can be used to represent the drift due to the creep effect.

Figure 2.38 shows the hysteresis loops resulting from Figure 2.37. The hysteresis loops also slowly drift owing to the creep effect. The hysteresis loops depart from the nominal loop as the sinusoidal period input voltage increases. The creep effect is contained in the hysteresis output. The coupling between the hysteresis and the creep effect will increase the difficulty of identification and compensation.

Moreover, simple compensation is presented in this section. The nominal positions of the circles in Figure 2.38 are zero. Thus, one curve-fitting technique is used to compensate for the creep effect. Linear approximations are used to compute the creep displacement between two neighboring circles. Figure 2.39 shows the creep compensation obtained with the curve-fitting technique.

Figure 2.40 shows the hysteresis loops after creep compensation. It can be seen that the drifts due to the creep effect have been compensated.

2.9 Modeling Suggestions for Systems with Hysteresis

The models of systems with hysteresis are complex. The modeling of systems with hysteresis can be divided into four types—that is, linear dynamics, static hysteresis, direct dynamic hysteresis,

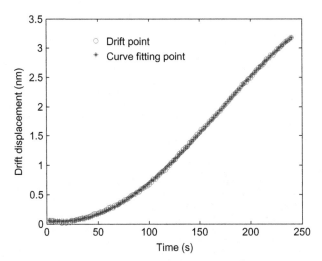

Figure 2.39 Piezoelectric displacement with drift suppression.

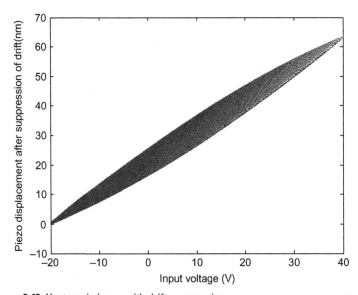

Figure 2.40 Hysteresis loops with drift suppression.

and composite hysteresis. This chapter also presents the typical model components of mechanical vibrations, electrical dynamics, static hysteresis, and the creep effect. This section will give some suggestions for how to model smart systems with hysteresis.

First, *linear dynamics*. The quasi-static hysteresis is regarded as a nominal gain k_h ($k_h = 1.02$ in the experiment) and an

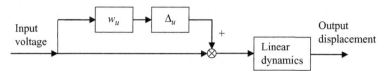

Figure 2.41 Input uncertainty of the linear dynamics due to the static hysteresis in smart systems.

input uncertainty of the nonhysteretic dynamics. The hysteresis contributes an input uncertainty to the nonhysteretic dynamics. Figure 2.41 illustrates the input uncertainty due to the rate-independent static hysteresis. The function w_u represent the hysteresis ratio in the input voltage. The complex uncertainty Δ_u has maximum norm $\|\Delta_u\| = 1$. In different smart systems, w_u is different. In piezoelectric systems with high stiffness, w_u is typically 5-15%, but in shape-memory-alloy systems, w_u can be as large as 30%.

Next, *static hysteresis*. At low frequencies, the response of smart systems is mainly contributed by the static hysteresis. In scanning or other smooth motion control at low frequencies, static hysteresis models such as the Preisach model and the Prandtl-Ishlinskii model can give satisfactory modeling accuracy. Ge and Jouaneh [1] used the classical Preisach hysteresis to model a piezoelectric actuator, and this gives an effective result. We also used the classical Preisach hysteresis to model and compensate the low frequency response of an piezoelectric stage. The static Preisach hysteresis improves the proportional-integral-derivative tracking performance by more than 80%.

Next, *direct dynamic hysteresis*. Direct dynamic hysteresis is modified from static hysteresis. Dynamic hysteresis can be used to model smart systems with hysteresis at broadband frequencies, but this approach is more suitable for periodic and smooth signals such as sinusoidal and triangle waves, and inversion-based feed-forward is a related compensation method. It is difficult to apply a general signal. Further, modern controllers are not suitable for direct dynamic hysteresis.

Finally, the *composite representation*. The composite representation of dynamic hysteresis is an alternative approach to describe the dynamic hysteresis. The composite representation is a synthesis of mechanical vibration, electrical dynamics, static hysteresis, and the creep effect. It is easier to understand and identify the composite representation. The composite representation is suitable for modern control design.

The model selection of smart systems with hysteresis is related to the modeling accuracy specification. In applications with modest accuracy, linear dynamics such as the mechanical vibration dynamics and capacitor dynamics are suitable. In applications with low-frequency scanning or smooth motion, the static hysteresis model can be used to represent smart systems. Moreover, the proportional-integral-derivative tuning can be done separately without considering the static hysteresis model. In applications with fast scanning or periodic smooth motion, direct dynamic hysteresis can be used. Dynamic hysteresis-based inversion-based feedforward control can be used to compensate the dynamic hysteresis.

In applications with both fast and precision scanning or motion control, the composite representation can be used to describe the dynamic hysteresis over a broad range of frequencies. The control system can also be designed according to the composite representation. The feedforward compensator can be designed with the composite representation. The feedback control can be designed with the linear components in the composite representation. Thus, various modern and intelligent control approaches can be developed with the composite representation.

2.10 Conclusions

The fundamentals of smart systems with hysteresis have been presented in this chapter. The mechanical vibration and capacitor dynamics (electrical dynamics) have been illustrated. The Preisach model, Preisach plane, and properties of the static hysteresis were investigated in detail. The behavior of the Preisach hysteresis and that of the linear dynamics phase delay were compared. Further, the closed-loop response of a smart system with hysteresis was presented. To model smart hysteresis at broadband frequencies, the dynamic hysteresis was investigated. Moreover, the composite representation is an alternative proposed to describe the dynamic hysteresis. Finally, the modeling suggestions were presented.

References

[1] P. Ge, M. Jouaneh, Tracking control of a piezoceramic actuator, IEEE Trans. Control Syst. Technol. 4 (3) (1996) 209-216.
[2] I.D. Mayergozy, Mathematical Modeling of Hysteresis, Springer-Verlag, New York, 1991.
[3] D. Davino, C. Natale, S. Pirozzi, C. Visone, Phenomenological dynamic model of a magnetostrictive actuator, Phys. B Cond. Matter Phys. 343 (2004) 112-116.
[4] A. Visintin, Differential Models of Hysteresis, Springer-Verlag, New York, 1994.

[5]   K. Kuhnen, Modeling, identification and compensation of complex hysteretic nonlinearities: a modified Prandtl-Ishlinkskii approach, Euro. J. Control 9 (2003) 407-418.

[6]   O. Henze, W.M. Rucker, Identification procedures of Preisach model, IEEE Trans. Magnet. 38 (2) (2002) 833-836.

[7]   M. Rakotondrabe, C. Clevy, P. Lutz, Hysteresis and vibration compensation in a nonlinear unimorph piezocantilever, in: IEEE/RSJ International Conference on Intelligent Robots and Systems, 2008, pp. 300-306.

[8]   C.J. Lin, S.R. Yang, Precise positioning of piezo-actuated stages using hysteresis observer based control, Mechatronics 16 (2006) 417-426.

[9]   I. Mayergoyz, Dynamic Preisach models of hysteresis, IEEE Trans. Magnet. 24 (6) (1988) 2925-2927.

[10] M.A. Janaideh, S. Rakheja, C.Y. Su, An analytical generalized Prandtl-Ishlinskii model inversion for hysteresis compensation in micropositioning control, IEEE/ASME Trans. Mechatron. 116 (4) (2011) 734-744.

[11] X. Tan, J.S. Baras, Adaptive identification and control of hysteresis in smart materials, IEEE Trans. Automat. Control 50 (6) (2005) 827-839.

3

HYSTERESIS MODELING IN SMART ACTUATORS

Abstract

Smart actuators are being widely applied with the rapid development of precision instruments. In this chapter, we mainly focus on actuators made with smart materials, such as piezoelectric materials, magnetostrictive materials, and shape-memory-alloy materials. The hysteresis modeling of smart hysteresis is investigated. The static hysteresis, composite hysteresis, and persistent-excitation problem are presented. Simple identification approaches are also provided. Finally, the response components of smart actuators are illustrated.

Keywords: Composite hysteresis, Smart actuator, Singular value decomposition, Creep, Mechanical vibration

Modeling and Precision Control of Systems with Hysteresis
http://dx.doi.org/10.1016/B978-0-12-803528-3.00003-3

3.1 Introduction

In this chapter, hysteresis modeling of smart actuators without voltage amplifier dynamics will be presented. Smart materials such as piezoceramic materials, shape-memory-alloy materials, and magnetostrictive materials are widely applied in precision engineering [1, 2]. Among the nonlinearities present in smart materials, hysteresis contributes a main uncertainty which affects control or measurement performance. In the open loop, the maximum error from hysteresis in piezoelectric actuators is 10-15% of the total displacement [3], which may not be tolerable for precision control applications.

Models of hysteresis can be classified into physical models [4, 5] and mathematical models. Physical models are constructed on the basis of physical laws applied to the phenomenon of hysteresis and thus models are intuitive but are typically in a complex form which is difficult to identify and use for control. Conversely, mathematical models are a vehicle to provide the input-output relationship of the actual system and are usually more amenable to practical use for identification and control. In the current literature, the ferromagnetic hysteresis model [4] is a physical model, and the Preisach model and the Prandtl-Ishlinskii model [6] are two popular mathematical models [4]. The Prandtl-Ishlinskii model has been used to compensate the hysteresis nonlinearity of piezoelectric actuators [7–9]. The Bouc-Wen model has also been applied to describe the nonlinearity of lead zirconate titanate (PZT) actuators [10–12], which exhibit hysteretic characteristics [7]. Compared with the Presiach model, these models are in a highly nonlinear form possibly including dynamic parameters which are difficult to represent in a form for identification and subsequent use for control purposes.

Modeling of Preisach hysteresis entails essentially the identification of Preisach density functions. Mayergoyz [4], Hu [13], Song et al. [14] identified these functions by differentiating the measurement data, causing the identified functions to be highly sensitive to measurement noise. Ge [3] and Jouaneh and Hu [13] studied the tracking control of PZT actuators using an off-line computation of the difference of measurement data. Iyer et al. [15] and Tan and Baras [16] developed recursive schemes for parameter identification and designed a closest-match algorithm for compensation of the Preisach hysteresis. Henze and Rucher [6] provided approaches for identification of the Preisach function based on different distribution characteristics. These approaches rely on assumptions of the form of the density functions, and they are typically not amenable for producing a model for hysteresis compensation and control. As evident in the published literature, the current approaches to identify the Preisach density functions in the continuous form are not practically viable in one way or another. Approximate Preisach density functions can be identified more efficiently through a discretized Preisach plane by transformation of the double integral of the density functions to a numerical summation. Furthermore, no restrictive assumption on the density functions is necessary.

With the models available, hysteresis compensators can be designed. The compensation of hysteresis through inverse of the models is adopted in some cases [14, 16]. Tao and Kokotovic [17] designed an adaptive controller with a parameterized inverse hysteresis, however with the hysteresis phenomenon simplified to a relay operation. Wu and Zhou [2] presented inversion-based iterative control of PZT actuators, using repetitive cycles to yield the indirect inverse model. Chen et al. [9, 18] developed adaptive techniques without requiring an inversion of the hysteresis, but the density function is still directly needed in the controller .

In this chapter, we use least squares to identify the density function for a large number of split lattices to obtain a smooth approximation of the continuous density functions and equivalently a large number of model parameters to be determined from the data. This leads to the requirement to collect a large amount of sufficiently data to satisfy a persistent-excitation (PE) condition [16, 19]. Under practical conditions of limited data collection, this issue is equivalent to solving an ill-conditioned inverse problem of a singular matrix as in the following operator equation [20]:

$$Ax = b, \tag{3.1}$$

where x and b belong to normed linear spaces, and A is an deficient matrix mapping x to b. In the least-squares sense, there will be infinite solutions to (3.1). However, a solution can be computed [20, 21] when a norm objective is set to minimize $\|Ax - b\|_2$ and $\|x\|_2$. In this chapter, we use a least-squares estimation algorithm with singular value decomposition (SVD) to identify linear dynamics with Preisach hysteresis, and then we present the linearization of hysteresis. Tracking of complex signals indicates that the linearization can improve the tracking performance.

3.2 Simplified Composite Representation of Smart Actuators

3.2.1 Static Preisach Hysteresis

The hysteresis relay constitutes the basic element, where the output $\gamma_{\alpha\beta}$ of the operator is assumed to be of two values (e.g., 0 and 1) [3, 13], where α and β are the switching threshold values of the hysteresis operator $\gamma_{\alpha\beta}[u(t)]$. The hysteresis output—that is, the voltage-to-displacement dynamics—is given in (3.2):

$$f(t) = \iint\limits_{\alpha \geq \beta} \mu(\alpha, \beta)\gamma_{\alpha\beta}[u(t)] \, d\alpha \, d\beta, \tag{3.2}$$

where $\mu(\alpha, \beta)$ is the distribution weight function and $f(t)$ is the hysteresis output. In the Preisach model, the same input $u(t)$ is applied to all the hysteresis operators $\gamma_{\alpha\beta}[u(t)]$. The output of these operators is weighted by the Preisach density function $\mu(\alpha, \beta)$ unique to the actuator, and then summed continuously over possible values of α and β. The hysteresis output can thus be considered to be a superposition of a continuous set of two-position relay operators $\gamma_{\alpha\beta}[u(t)]$ over the range of the input signal. From this way of interpreting the input-output relationship, the classical Preisach model in (3.2) can be rewritten as (3.3), which is a weighted computation of the area of the limiting region thus evolved:

$$f(t) = \iint\limits_{S} \mu(\alpha, \beta)\gamma_{\alpha\beta}[u(t)] \, d\alpha \, d\beta, \tag{3.3}$$

where S is the Preisach triangle which is formed by $\alpha \geq \beta$ and the saturation value of input voltages, and can be divided into S^+, with $\gamma_{\alpha\beta}[u(t)] = \theta_1$, and S^-, with $\gamma_{\alpha\beta}[u(t)] = \theta_2$. Thus, (3.2) can be written as

$$f(t) = \iint\limits_{S^+} \mu(\alpha, \beta)\gamma_{\alpha\beta}[u(t)] \, d\alpha \, d\beta + \iint\limits_{S^-} \mu(\alpha, \beta)\gamma_{\alpha\beta}[u(t)] \, d\alpha \, d\beta$$

$$= \theta_1 \iint\limits_{S^+} \mu(\alpha, \beta) \, d\alpha \, d\beta + \theta_2 \iint\limits_{S^-} \mu(\alpha, \beta) \, d\alpha \, d\beta$$

$$= (\theta_1 - \theta_2) \iint\limits_{S+} \mu(\alpha, \beta) \, d\alpha \, d\beta + \theta_2 \iint\limits_{S} \mu(\alpha, \beta) \, d\alpha \, d\beta$$

$$= (\theta_1 - \theta_2) \iint\limits_{S+} \mu(\alpha, \beta) \, d\alpha \, d\beta + c\theta_2, \tag{3.4}$$

where $c = \iint\limits_{S} \mu(\alpha, \beta) \, d\alpha \, d\beta$ is a constant.

For example, in piezoelectric actuators, it can be assumed that $\theta_1 = 1$ and $\theta_2 = 0$, and (3.4) is thus rewritten as (3.5):

$$f(t) = \iint\limits_{S+} \mu(\alpha, \beta) \gamma_{\alpha\beta} [u(t)] \, d\alpha \, d\beta. \tag{3.5}$$

3.2.2 Simplified Composite Representation

In this section, the smart system is represented by a simplified composite representation, as shown in Figure 3.1. The linear dynamics component is represented by a second-order model incorporating input Preisach hysteresis. Further, a linear time invariant model that is strictly rational is considered, and it can be represented by the transfer function as follows:

$$G(s) = \frac{\overline{a}_{k-1} s^{k-1} + \overline{a}_{k-2} s^{k-2} + \cdots + \overline{a}_0}{s^n + \overline{b}_{k-1} s^{k-1} + \overline{b}_{k-2} s^{k-2} + \cdots + \overline{b}_0}. \tag{3.6}$$

As $G(s)$ is strictly rational, $G(s) \mapsto \overline{a}_0 / \overline{b}_0$ when $s \mapsto 0$. $\overline{a}_0 / \overline{b}_0$ is also called the DC gain of smart systems. Without loss of generality, a second-order plant with damping ratio 0.2 and resonant frequency 50 rad/s is used in this chapter, as shown in (3.7):

$$G(s) = \frac{\omega^2}{s^2 + 2\xi\omega s + \omega^2}, \tag{3.7}$$

where s is the Laplace operate, ξ is the damping ratio, and ω is the resonant frequency.

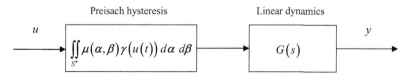

Figure 3.1 Linear dynamics with input hysteresis.

The Preisach loop between the input $u(t)$ and hysteresis output $f(t)$ is shown in Figure 3.2, in which the density function $\mu(\alpha, \beta)$ is set to 4 and the input signal is set to $2.25(\sin(15t + 1) + \sin(30t - 0.2) + \sin(75t) + 2)$. After linear dynamics, the Preisach hysteresis loop is distorted by linear dynamics, as shown in Figure 3.3 the whole system exhibits complex dynamical behavior.

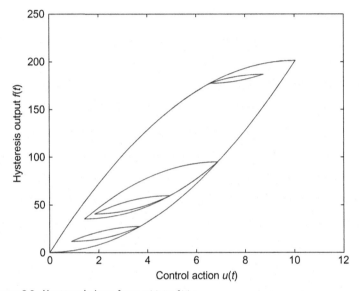

Figure 3.2 Hysteresis loop from $u(t)$ to $f(t)$.

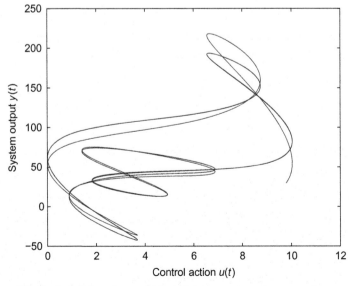

Figure 3.3 Hysteresis loop from $u(t)$ to $y(t)$.

Table 3.1 Convergence Values for Different Proportional and Integral Values

Proportional	Integral	Convergence Values
0.01	0.1	0.705
0.05	0.5	0.705, 0.41
0.1	1	0.705, 0.334
0.2	2	0.705, 0.26
0.3	3	0.705, 0.267
0.4	4	0.705, 0.514
0.5	5	0.705, 0.544

3.2.3   Closed-loop Control Property

For a step reference, the control input $u(t)$ converges to different values with different proportional and integral values, as shown in Table 3.1, and the control $u(t)$ can switch between several values, as shown in Figure 3.4, although the same steady output can be obtained.

3.2.4   Simplified Identification Approach for Composite Hysteresis

Hysteresis Measurable

First, the hysteresis output is assumed to be measurable. In this section we investigate the identification of hysteresis systems by least-squares estimation using SVD. In this situation, the identification of the Preisach model essentially entails the identification of the density function $\mu(\alpha, \beta)$. In the continuous form, the parameter $\mu(\alpha, \beta)$ is continuous over the limiting region, and it is thus difficult to directly identify the continuous Preisach functions. Instead, the limiting region will be considered as comprising discrete lattices. Each lattice cell has a weight assigned to it, which is the discrete equivalent of a specific lattice density. s_i denotes the area of lattice i.

Then, the hysteresis output can be computed by transformation of the double integral to a numerical summation as shown

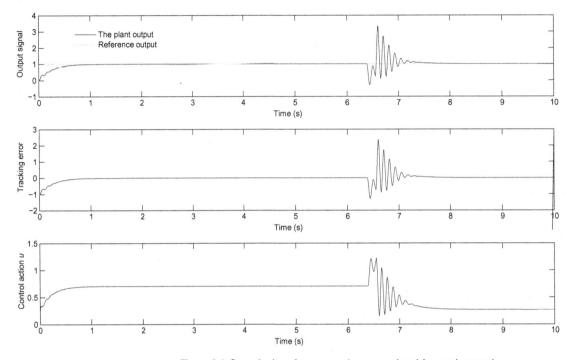

Figure 3.4 Step single reference under proportional-integral control.

in (3.8), which is also linear in parameter and suitable for parameter estimation:

$$f(t) = \iint\limits_{S^+} \mu(\alpha, \beta)\gamma_{\alpha\beta}[u(t)] \, d\alpha \, d\beta$$

$$= \sum\sum \gamma_{ij}[u(t)]\mu_{ij}s_{ij} + o(t), \tag{3.8}$$

where $o(t)$ is the error from discretization. The estimation of hysteresis output can be represented in matrix form as

$$\hat{f}(t) = R^{\mathrm{T}}(t)V. \tag{3.9}$$

The Preisach plane is discretized into $L \times L$ lattices; thus, effectively there are $L^2/2$ lattices in the Preisach triangle to be identified. The PE problem is associated with the large number of parameters and the limited frequencies in the input signal, particularly when data are collected with the system under closed-loop control. This PE condition is mathematically equivalent to an ill-conditioned matrix which can be addressed by incorporation of SVD in the least-squares estimation.

Over a time range $t_0 < t < t_N$, data samples are collected at times $t_0, t_1, \ldots, t_i, \ldots, t_N$. With N samplings, (3.1) can be produced by using equation (3.9), where A is represented as

$$
A = \begin{bmatrix}
\gamma_{11}^0 & \gamma_{12}^0 & \cdots & \gamma_{1L}^0 & \gamma_{22}^0 & \gamma_{23}^0 & \cdots & \gamma_{2L}^0 & \cdots & \gamma_{LL}^0 \\
\gamma_{11}^1 & \gamma_{12}^1 & \cdots & \gamma_{1L}^1 & \gamma_{22}^1 & \gamma_{23}^1 & \cdots & \gamma_{2L}^1 & \cdots & \gamma_{LL}^1 \\
\vdots & \vdots & \cdots & \vdots & \vdots & \vdots & \cdots & \vdots & \cdots & \vdots \\
\gamma_{11}^N & \gamma_{12}^N & \cdots & \gamma_{1L}^N & \gamma_{22}^N & \gamma_{23}^N & \cdots & \gamma_{2L}^N & \cdots & \gamma_{LL}^N
\end{bmatrix}.
$$

A is $L(L+1)/2 \times (N+1)$, $x = V$, and b is written as

$$
b = \begin{bmatrix}
f(0) \\
f(1) \\
\vdots \\
f(N)
\end{bmatrix}.
$$

Let $\hat{x}$ be the estimate of x, and let E be the error between $A\hat{x}$ and b as in (3.10):

$$
E = A\hat{x} - b. \tag{3.10}
$$

Case 1   ($M = A^{\mathrm{T}}A$ is nonsingular; Theorem (least-squares estimation) [19]):
The norm of error $\|Ax - b\|$ can be minimal for parameter x, and also the minimum is unique and is given by

$$
\hat{x} = (A^{\mathrm{T}}A)^{-1}A^{\mathrm{T}}b. \tag{3.11}
$$

Proof.

$$
\begin{aligned}
\|E\|_2^2 &= E^{\mathrm{T}}E \\
&= (A\hat{x} - b)^{\mathrm{T}}(A\hat{x} - b) \\
&= b^{\mathrm{T}}A\hat{x} - b^{\mathrm{T}}b - \hat{x}^{\mathrm{T}}A^{\mathrm{T}}A\hat{x} + \hat{x}^{\mathrm{T}}A^{\mathrm{T}}b \\
&= 2\hat{x}^{\mathrm{T}}A^A b - \hat{x}^{\mathrm{T}}A^{\mathrm{T}}A\hat{x} - b^{\mathrm{T}}b, \tag{3.12}
\end{aligned}
$$

$$
\frac{\partial \|E\|_2^2}{\partial x} = 0. \tag{3.13}
$$

We can obtain

$$2A^T b - 2A^T A\hat{x} = 0. \tag{3.14}$$

Since $A^T A$ is nonsingular or of full rank, the least-squares estimation of x is

$$x = (A^T A)^{-1} A^T b. \tag{3.15}$$

$\square$

Case 2   ($M = A^T A$ is singular and ill conditioned): If $A^T A$ is singular—that is, $\det(A^T A) = 0$, the PE condition is not satisfied, which is likely to occur in this application of Preisach modeling. In this case, SVD can be used to obtain the pseudoinverse of $A^T A$ instead. The SVD of $A^T A$ is given as

$$M = U\Sigma V, \tag{3.16}$$

where $U = \begin{bmatrix} u_1 & u_2 & \cdots & u_n \end{bmatrix}$,
$V = \begin{bmatrix} v_1 & v_2 & \cdots & v_n \end{bmatrix}$, and $\Sigma = \mathrm{diag}(\sigma_1, \sigma_2, \ldots, \sigma_n)$
$\sigma_1 \geq \sigma_2 \geq \cdots \geq \sigma_n$, where U and V are unitary matrices satisfying the conditions below:

$$\begin{aligned} UU^T &= U^T U = I, \\ VV^T &= V^T V = I, \end{aligned} \tag{3.17}$$

where $U^{-1} = U^T$ and $V^{-1} = V^T$.
Assume the rank of matrix M is k, then $\sigma_{k+1}, \sigma_{k+2}, \ldots, \sigma_n = 0$ if $\frac{\sigma_j}{\sigma_1} \ll 1, j = r+1, r+2, \ldots, k$. M can be decomposed as follows:

$$M = \begin{bmatrix} U_r & U_{n-r} \end{bmatrix} \begin{bmatrix} \Sigma_r & 0 \\ 0 & \Sigma_{n-r} \end{bmatrix} \begin{bmatrix} V_r & V_{n-r} \end{bmatrix}, \tag{3.18}$$

where $U_r = \begin{bmatrix} u_1 & u_2 & \cdots & u_r \end{bmatrix}$, $U_{n-r} = \begin{bmatrix} u_{r+1} & u_{r+2} & \cdots & u_n \end{bmatrix}$,
$V_r = \begin{bmatrix} v_1 & v_2 & \cdots & v_r \end{bmatrix}$, $V_{n-r} = \begin{bmatrix} v_{r+1} & v_{r+2} & \cdots & v_n \end{bmatrix}$,
$\Sigma_r = \mathrm{diag}([\sigma_1, \sigma_2, \ldots, \sigma_r])$, and $\Sigma_{n-r} = \mathrm{diag}([\sigma_{r+1}, \sigma_{r+2}, \ldots, \sigma_n])$,
and where U_r, U_{n-r}, V_r, and V_{n-r} are orthonormal matrices. Then, it follows that

$$M = \sum_{i=1}^{k} \sigma_i u_i v_i^T$$

$$= \sum_{1}^{r} u_i v_i^{\mathrm{T}} + \sum_{r+1}^{n} u_i v_i^{\mathrm{T}}, \qquad (3.19)$$

where $\| \sum_{r+1}^{n} u_i v_i^{\mathrm{T}} \|_{L_2} = \sigma_{r+1}$.

Small singular values can be truncated to obtain a better least-squares estimation in a ill-conditioned situation while preserving the main "energy" of M. The approximation of the pseudoinverse of M is given by

$$M^+ \approx V \Sigma^{-1} U^{\mathrm{T}} A^{\mathrm{T}}, \qquad (3.20)$$

where

$$\Sigma^{-1} = \mathrm{diag}([\sigma_1, \sigma_2, \dots, \sigma_r, 0, \dots, 0]). \qquad (3.21)$$

The pseudoinverse can be reformulated as follows:

$$M^+ = \sum_{1}^{r} \frac{1}{\sigma_i} v_i u_i^{\mathrm{T}} A^{\mathrm{T}}. \qquad (3.22)$$

When $\sigma_i \to 0$, $M^+ \to \infty$. Hence, minimizing $\|x\|$ is necessary, and small σ_i needs to be truncated. Then, we can obtain the estimation of x in the least-squares sense:

$$\hat{x} = M^+ b. \qquad (3.23)$$

In simulation, $\alpha_{\max}$ is set to 10, $\alpha_{\min}$ is set to 0, $\beta_{\max}$ is set to 10, $\beta_{\min}$ is set to 0, and $\mu(\alpha, \beta)$ is set to 4 (uniform lattice density). With 4000 sampling points, A is a 4000×1275 matrix.

To describe the identification accuracy, $\|x - \hat{x}\|_2$ and $\|x\|_2$ are used as shown in (3.24) and (3.25). Then, we can get $\|x\|_2 = 140.47$. If the determinant $\det(M) = 0$ (i.e., M is singular) and the rank is 99, and if we let $r = 64$ with truncation $\sigma_i/\sigma_1 = 0.001$, the estimation error $\|x - \hat{x}\|_2 = 6.67$:

$$\|x - \hat{x}\| = \sqrt{(\mu_1 - \hat{\mu}_1)^2 + \cdots + (\mu_{L(L+1)/2} - \hat{\mu}_{L(L+1)/2})^2}, \qquad (3.24)$$

$$\|x\|_2 = \sqrt{\mu_1^2 + \mu_2^2 + \cdots + \mu_{L(L+1)/2}^2}. \qquad (3.25)$$

Estimates of $\mu(\alpha, \beta)$ and the estimation error are shown in Figure 3.5, from which it can be seen that the difference between the real and estimated density distribution function $\mu(\alpha, \beta)$ is small, $\|x - \hat{x}\|/\|x\| = 4.7\%$ with 4000 samples.

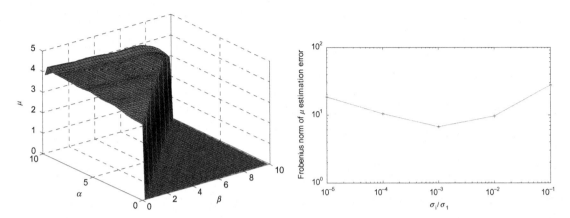

Figure 3.5 Parameter identification result with truncation $\sigma_r/\sigma_1 = 0.001$ and the different errors from different truncations.

The estimation of $\mu(\alpha, \beta)$ is shown in Figure 3.5. It also can be seen that the estimated errors are different for different truncation thresholds. This indicates that a larger truncation threshold increases the approximation error of matrix M by suppressing the ill-conditioned problem in the pseudoinverse; however, a smaller truncation threshold increases the error of the inverse, where the ill-conditioned problem becomes more serious, especially if there is noise in the system. The truncation of the singular value is a trade-off between minimizing the truncation error and minimizing $\|M^+\|$.

In the simulation in this chapter, the density function of Preisach hysteresis is set to a uniform value of 4. The PE condition problem is thus not serious. The identification of Preisach hysteresis can be finished by means of simple sinusoidal inputs. However, the density function of Preisach hysteresis is complex in reality, and the identification of Preisach hysteresis will be investigated with complex inputs in Section 4.4.3.

Hysteresis Unmeasurable

In reality, the hysteresis output is not measurable. In this situation, the Tustin transform in (3.26) can be used to transform the continuous transfer function $G(s)$ into the discrete one $G(z)$:

$$s = \frac{2}{T}\frac{z-1}{z+1},$$ (3.26)

$$G(z) = k_z \frac{z^2 + 2z + 1}{z^2 + b_1 z + b_0}.$$ (3.27)

As shown in (3.7), the DC gain $k = \overline{a}_0/\overline{b}_0 = 1$ of linear dynamics is absorbed by Preisach hysteresis, such that the whole system is linear in parameter, as shown in Figure 3.6. However, the density function of Preisach hysteresis can be identified with the factor k, which cannot be separated.

The simplified identification strategy is proposed as follows: First, an input signal of low frequency can be chosen which is much smaller than the second-order dynamics, whose resonance is given as 50 rad/s, where the contribution of second-order dynamics is small except for the DC gain and the phase degree at low frequency. The static hysteresis, which is independent of frequency, can be identified. Second, consider a high-frequency signal of 60 rad/s. We can eliminated the contribution from hysteresis by using the identification result for hysteresis. The matrix for the parameter identification of linear dynamics is reduced to a matrix with low dimension. When the condition number of the matrix is not too large (i.e., the matrix is not nearly singular), the exact estimation of the parameters of linear model can be given.

Step 1: The identification of classical Preisach hysteresis, which is rate independent, is presented with use of a low-

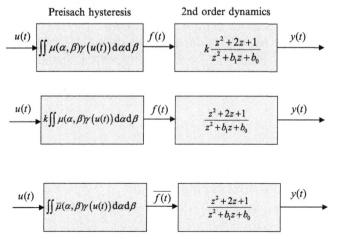

Figure 3.6 The identification process for hysteretic linear dynamics.

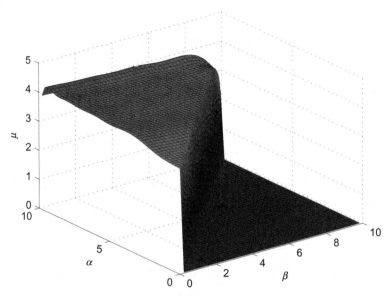

Figure 3.7 Identification of static Preisach hysteresis.

frequency signal where the DC gain and a very small phase degree are contributed by the linear dynamics. The sinusoidal $u(t) = 10|\sin(0.5t)|$ is used, and the output of the plant can be regarded as the corresponding hysteresis output with an uncertainty which contributes to the estimation error. Then, least-squares estimation by SVD is applied. The identification is shown in Figure 3.7 with the estimation error $\|x-\hat{x}\| = 6.9$, which is larger than in the case when the hysteresis output can be measured. However, the coefficient $k = 1$ in this chapter is a special case. In general, the density function of hysteresis can be identified with the coefficient k, which cannot be separated.

Step 2: The identification of linear dynamics with use of a high-frequency input can be done after step 1. Equation (3.28) represents the input-output relationship from hysteresis output $f(t)$ to system output $y(t)$:

$$\overline{f(t)} + 2\overline{f(t - 2T_\mathrm{s})} + \overline{f(t - T_\mathrm{s})} = y(t) + b_1 y(t - T_\mathrm{s}) + b_0 y(t - 2T_\mathrm{s}). \tag{3.28}$$

However, (3.9) represents the relationship between the control input u and the hysteresis output $f(t)$.

Let $x = \begin{bmatrix} b_0 & b_1 \end{bmatrix}^\mathrm{T}, a(t) = \begin{bmatrix} -y(t - 2T_\mathrm{s}) & -y(t - T_\mathrm{s}) \end{bmatrix}$,

$$\overline{y(t)} = y(t) - \overline{f(t)} + 2\overline{f(t - 2T_\mathrm{s})} + \overline{f(t - T_\mathrm{s})}$$

$$=y(t) - [R(t) + 2R(t - T_s) + R(t - 2T_s)]\overline{V},$$

where $\overline{V}$ is the estimate of V with the coefficient k from linear dynamics. Then, (3.29) can be given to estimate the parameters of the linear model:

$$a(t)x = \overline{y(t)}, \tag{3.29}$$

where x is the estimation vector. Take n sequences of (3.29), and let $A = \left[a(0)^{\mathrm{T}}, a(1)^{\mathrm{T}}, \ldots, a(n-1)^{\mathrm{T}}\right]^{\mathrm{T}}$ and $b = \left[\overline{y(0)}, \overline{y(1)}, \ldots, \overline{y(n-1)}\right]^{\mathrm{T}}$, then (3.1) for identification can be obtained. The PE condition for a second-order system is easy to satisfy; $u(t) = 10|\sin(60t)|$ is used, and the estimation values of $b_0 = 0.98$ and $b_1 = -1.9775$ are 0.9805 and -1.9780, respectively.

3.2.5 Simplified Control of Linear Dynamics with Input Static Preisach Hysteresis

With the parameters obtained via the possible approaches presented in previous section, the inversion-based linearization of hysteresis can be designed, as shown in Figure 3.8. At time instant k, define the reference control as $u_r(k)$, hysteresis output as $f(k)$, and control action as $u(k)$. The Preisach inverse linearization works as follows to obtain $u(k+1)$ based on memory curve Ψ and the estimation of parameter V. The process can be represented in the following pseudocode chart:

$$
\begin{aligned}
&\text{if} \quad \gamma_1 u_r(k) + \gamma_2 < \hat{f}(k) \\
&\quad u_{k+1} = u(k) - i\delta, \quad i = 1, \ldots, m \\
&\quad \hat{f}(k+1) = \Psi_1(u(k+1)) \\
&\text{if} \quad \gamma_1 u_r(k) + \gamma_2 \geq \hat{f}(k+1), \text{ break} \\
&\text{elseif} \quad \gamma_1 u_r(k) + \gamma_2 > \hat{f}(k)
\end{aligned}
$$

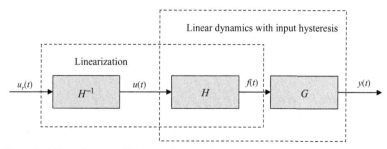

Figure 3.8 Linearization of Preisach hysteresis.

$$u_{k+1} = u(k) + i\delta, \quad i = 1, \dots, m$$
$$\hat{f}(k+1) = \Psi_2(u(k+1))$$
$$\text{if} \quad \gamma_1 u_r(k) + \gamma_2 \leq \hat{f}(k+1) \text{ break}$$
$$\text{else} \quad u(k+1) = u(k), \text{ end}$$

The iteration step δ is expressed as

$$\delta = \lambda \frac{u_{\max}}{L}, \tag{3.30}$$

where λ is a factor to regulate the step, and L is the discretization level of the Preisach plane. Control action $u(k+1)$ is computed to enforce $f(k+1)$ to track $\gamma_1 u_r(k) + \gamma_2$. Using the linearization, we can transform the hysteresis as a constant parallel with the residual error as shown in Figure 3.9.

Let $\gamma_1 = 20$ and $\gamma_2 = 0$ and apply linearization. It can be seen in Figure 3.10 that the hysteresis from reference control u_r to hysteresis output $f(t)$ is suppressed, compared with the hysteresis from the real control input $u(t)$ to hysteresis output $f(t)$. Finally, one proportional-integral-derivative (PID) controller with proportional $k_P = 4$, integral $k_I = 20$, and derivative $k_D = 0$ is

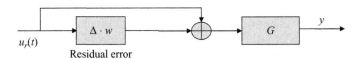

Figure 3.9 Linearization of Preisach hysteresis.

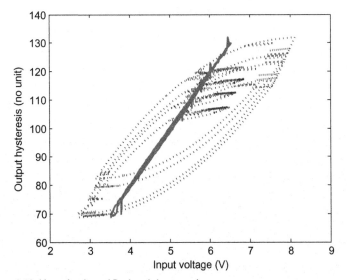

Figure 3.10 Linearization of Preisach hysteresis.

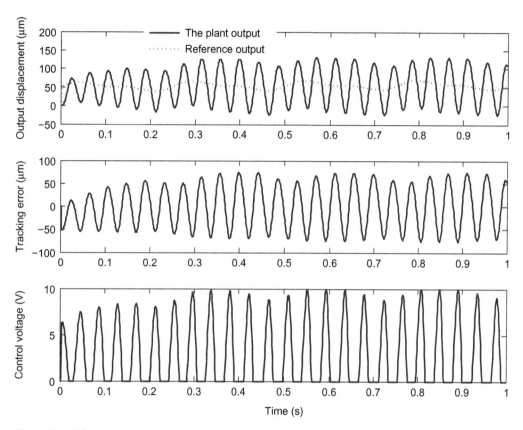

Figure 3.11 PID control without linearization.

applied to the linear hysteretic dynamics to track the reference signal $10 \sin t + 50 \sin(25t) + 5 \sin(50t) + 50$. The tracking result without linearization is shown Figure 3.11, which indicates that the tracking is not satisfied. The same PID controller with linearization is shown in Figure 3.12, and the tracking performance is better.

3.2.6 Persistent-Excitation Problem

When the density function is distributed in the Preisach plane, the PE problem is more difficult. It is desired to design the input voltage which has adequate amplitudes rather than frequencies. The sampling points are selected uniform in voltage rather than uniform in time. Figure 3.13 illustrates the designed input voltage and the sampling points. Figure 3.14 illustrates the sampling points in the hysteresis loops.

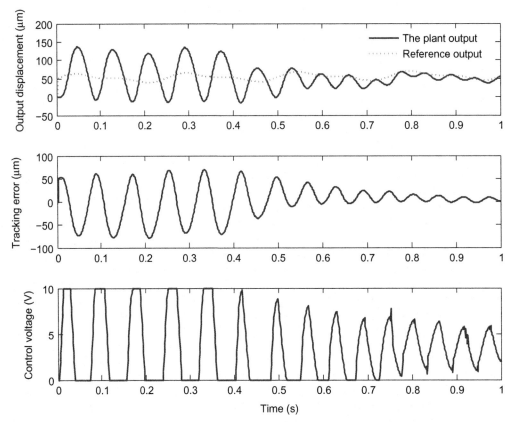

Figure 3.12 PID control with linearization.

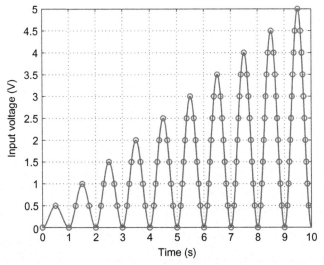

Figure 3.13 Sinusoidal inputs to avoid the PE problem (circles are sampling points).

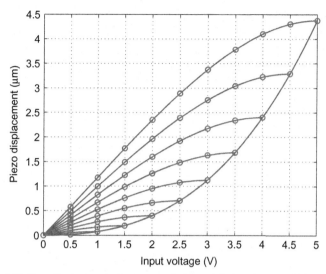

Figure 3.14 Sampling points in the hysteresis loops (circles are sampling points).

The input signal can be represented as

$$u(t) = \frac{P(t)}{2}(1 - \cos(\omega_r t)),\tag{3.31}$$

where t is time, $\omega_r = 1/T_{sub}$, where T_{sub} is the period of each harmonic signal, and $P(t)$ is the stair amplitude of the input signal represented in (3.32):

$$P(t) = \text{fix}\left(1 + \frac{t - \text{fix}(t/T)T}{T_{sub}}\right) \cdot \delta_l,\tag{3.32}$$

where fix(x) is a function to round x to its nearest integer toward zero and δ_l is given as

$$\delta_l = u_{max}/L,\tag{3.33}$$

where T is the whole period, L is the discretization level, and u_{max} is the maximum voltage in the identification.

3.3 Creep Effect

The creep effect in smart actuators is presented in this section. Compared with the static hysteresis, mechanical vibration, and electrical dynamics, the creep is a slow effect. We use linear dynamics to represent the creep effect. At low frequencies, the mechanical vibration and electrical dynamics are coupled.

3.3.1 Linear Creep Model

Linear dynamics can be represented as [22]

$$G_c(s) = \frac{1}{k_0} + \Sigma_{i=1}^N \frac{1}{c_i s + k_i}, \tag{3.34}$$

where k_0 denotes the elastic behavior, and k_i and c_i denote the spring constant and the damping constant. In smart systems, it is reported that a linear model order of 3 is acceptable to describe the creep effect [23]. Alternatively, a nonlinear model can also be used to represent the creep effect, such as the following equation [24]:

$$y_c(t) = y_{c0}\left(1 + \gamma_c \log \frac{t}{t_0}\right), \tag{3.35}$$

where y_c is the displacement at time t, y_{c0} is the displacement at time t_0, and γ_c is the creep rate.

3.3.2 Coupled Hysteresis and Creep Effects

The static hysteresis effect is typically coupled with the creep effect. The static hysteresis is rate independent. It is a strong nonlinearity and has global memories. Compared with the rate-independent static hysteresis effect, the creep effect has slow dynamics. The creep behaves as drifts and degrades the identifi-

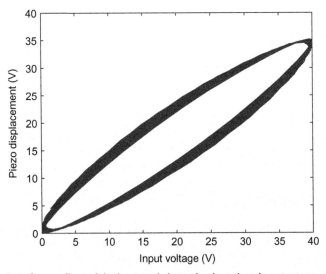

Figure 3.15 Creep effect of the hysteresis loops in piezoelectric actuators.

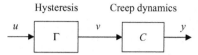

Figure 3.16 Coupled hysteresis and creep.

cation accuracy of the hysteresis model. Figure 3.15 illustrates the observed hysteresis loop in our experimental piezoelectric stage. The input voltage is a sinusoidal wave with an amplitude of 40 V. The hysteresis loops are distorted by the creep effect. Obvious coupling between the hysteresis and the creep is observed.

Figure 3.16 shows the series connection of the static Preisach hysteresis and the creep effect. Γ denotes the static Preisach hysteresis. C denotes the creep effect, and u, v, and y denote the input voltage, hysteresis output, and creep output, respectively.

3.4 Mechanical Vibration and the RC Effect in Piezoelectric Actuators

Mechanical vibrations and electrical dynamics were presented in Section 2.3. In this section were further present the synthesis of the mechanical vibration and electrical dynamics, as shown in Figure 3.17. T_{em} denotes the inverse piezoelectricity, in which the electrical energy is transformed to mechanical motion. The deduced force F drives the piezoelectric actuator to exhibit displacement x.

The transfer function from the input voltage u to the displacement x is written as

$$\frac{X(s)}{U(s)} = \frac{T_{em}}{RCs + 1} \frac{1}{ms^2 + cs + k},\tag{3.36}$$

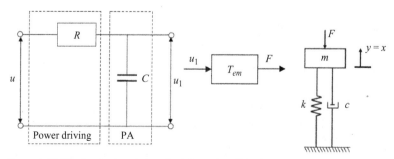

Figure 3.17 Mechanical and electrical dynamics. PA, piezoelectric actuator.

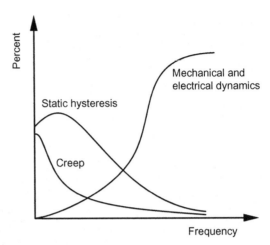

Figure 3.18 Dynamics.

where RC denotes the time constant of the capacitor dynamics, and m, c, and k represent the equivalent mass, damping, and stiffness, respectively.

3.5 Dynamics and Effects of Smart Actuators at Different Frequencies

Smart systems have static hysteresis, mechanical vibration, electrical dynamics, and a creep effect. Figure 3.18 gives a piezo-electric example of the responses of the dynamics and effects at different frequencies. The creep typically exhibits slow dynamics. As the input frequency increases, the creep percentage decreases. The static hysteresis is stronger than the creep effect. As the input frequency increases, the static hysteresis percentage decreases more slowly. The mechanical and electrical dynamics are fast dynamics. As the input frequency increases, the mechanical vibration percentage and the electrical dynamics percentage also increase. At low frequencies, the responses of smart systems are contributed mainly by the static hysteresis and the creep effect. At high frequencies, the responses of smart systems are contributed mainly by the mechanical vibration and electrical dynamics.

References

[1] W.M. Kuo, S.F. Chuang, C.Y. Nian, Y.S. Tarng, Precision nano-alignment system using machine vision with motion controlled by piezoelectric motor, Mechatronics 18 (1) (2008) 21-34.

[2] Y. Wu, Q. Zou, Iterative control approach to compensate for both the hysteresis and the dynamics effects of piezo actuators, IEEE Trans. Control Syst. Technol. 15 (5) (936-944) 2007.

[3]   P. Ge, M. Jouaneh, Tracking control of a piezoceramic actuator, IEEE Trans. Control Syst. Tech. 4 (3) (1996) 209-216.

[4]   I.D. Mayergozy, Mathematical Modeling of Hysteresis and Their Application, Elsevier, Amsterdam, 1991.

[5]   D. Davino, C. Natale, S. Pirozzi, C. Visone, Phenomenological dynamic model of a magnetostrictive actuator, Phys. B Cond. Matter Phys. 343 (1) (2004) 112-116.

[6]   O. Henze, W.M. Rucker Identification procedures of Preisach model, IEEE Trans. Magnet. 38 (2) (2002) 833-836.

[7]   M. Rakotondrabe, C. Clevy, P. Lutz, Hysteresis and vibration compensation in a nonlinear unimorph piezocantilever, in: International Conference on Intelligent Robots and Systems, IEEE/RSJ, Nice, 2008.

[8]   M. Janaiden, C. Su, S. Rakheja, Development of the rate-dependent Prandtl-Ishlinskii model for smart actuators, Smart Mater. Struct. 17 (2008) 035026.

[9]   X. Chen, T. Su, T. Fukuda, Adaptive control for the systems preceded by hysteresis, IEEE Trans. Automat. Control 53 (4) (2008) 1019-1025.

[10] J.C. Lin, S.R. Yang, Precise positioning of piezo-actuated stages using hysteresis observer based control, Mechatronics 16 (7) (2006) 417-426.

[11] A. Putra, S. Huang, K. Tan, Design, modeling, and control of piezoelectric actuators for intracytoplasmic sperm injection, IEEE Trans. Control Syst. Tech. 15 (2) (2007) 879-890.

[12] F. Ikhouane, J. Rodellar, A linear controller for hysteresis systems, IEEE Trans. Automat. Control 51 (2) (2006) 340-344.

[13] H. Hu, Compensation of hysteresis in piezoceramic actuators and control of nano-positioning system, Ph.D. dissertation, University of Toronto, Toronto, 2003.

[14] G. Song, J. Zhao, X. Zhou, J. Abreu-Garcia, Tracking control of a piezoceramic actuator with hysteresis compensation using inverse Preisach model, IEEE Trans. Magnet. 10 (2) (2005) 198-209.

[15] R. Iyer, X. Tan, P. Krishnaprasad, Approximate inversion of Preisach hysteresis operator with application to control of smart actuators, IEEE Trans. Automat. Control 50 (6) (2005) 798-809.

[16] X. Tan, J. Baras, Adaptive identification and control of hysteresis in smart materials, IEEE Trans. Automat. Control 50 (6) (2005) 827-839.

[17] G. Tao, P. Kokotovic, Adaptive control of plants with unknown hysteresis, IEEE Trans. Automat. Control 40 (2) (1995) 200-212.

[18] X. Chen, T. Hisayama, C. Su Pseudo-inverse-based adaptive control for uncertain discrete time systems preceded by hysteresis, Automatica 45 (2) (2009) 469-476.

[19] K. Astrom, B. Wittenmark, Adaptive control, 2nd ed., Addison-Wesley, New York, 1994.

[20] L. Lebedev, I. Vorovich, G. Gladwell, Functional Analysis Applications in Mechanics and Inverse Problem, Kluwer, New York, 2002.

[21] G. Golub, C. Loan, Matrix Computations, Johns Hopkins University Press, Baltimore, 1996.

[22] L.E. Malvern (Ed.), Introduction to the Mechanics of a Continuous Medium, Prentice Hall, Englewood Cliffs, NJ, 1969, pp. 313-319.

[23] D. Croft, G. Shed, S. Devasia, Creep, hysteresis, and vibration compensation for piezoactuators: atomic force microscopy application, ASME J. Dynam. Syst. Measur. Control 123 (1) (2001) 35-43.

[24] H. Jung, J.Y. Shim D. Gweon New open-loop actuating method of piezoelectric actuators for removing hysteresis and creep, Rev. Sci. Instr. 71 (9) (2000) 3436-3440.

4

COMPREHENSIVE MODELING OF MULTIFIELD HYSTERETIC DYNAMICS

Abstract

Piezoelectric systems are the most popular systems among smart systems. In this chapter we investigate the hysteresis modeling, identification, and precision control of smart systems where multifield effects and dynamics are considered by using two different piezoelectric smart mechanisms.

Modeling and Precision Control of Systems with Hysteresis
http://dx.doi.org/10.1016/B978-0-12-803528-3.00004-5

Keywords: Multi field hysteretic dynamics, Identification, Soft piezoelectric mechanism, Hard piezoelectric mechanism

First, the system description and experimental setup of typical piezoelectric smart systems are proposed. Next, the hysteretic dynamics is completely modeled, in which the static Preisach hysteresis and creep, electrical dynamics, and vibration dynamics are derived from the material, electrical, and mechanical fields, respectively. Then, according to the model characteristics, we present a comprehensive identification approach. A novel technique is provided to identify the electrical and vibration dynamics. Special inputs and sampling are proposed to identify the Preisach hysteresis. An experimental study is provided to demonstrate the effectiveness of the proposed modeling and identification approaches. Further, the control of the piezoelectric mechanism with high stiffness is also proposed. The modeling, identification, and control approaches in this chapter will be beneficial to further developments and high-performance motion of flexure-guided piezoelectric systems.

4.1 Introduction

Smart mechanisms, especially piezoelectric systems, are becoming more and more appealing in precision engineering [1–4], but the control performance is limited by the hysteretic dynamics, which belongs to the multifield and multiple-timescale domain. For a flexure-guided piezoelectric smart platform, the hysteretic dynamics is more complex. To enhance the control performance of the flexure-guided piezoelectric smart stage which can be used in nanopositioning and intracytoplasmic sperm injection [3], the accurate modeling, identification, and control will be investigated in this chapter. Flexible joints and motion-amplifying arms are often designed in piezoelectric-based platforms. The flexible joints are used to avoid friction and clearance nonlinearity. In this chapter, to achieve a larger displacement with a lower driving voltage, we use low-stiffness flexible joints instead of high-stiffness flexible joints. The proposed flexure-guided platform thus has lower resonant frequencies, which increases the difficulty of the modeling and identification of the hysteretic dynamics. Additionally, to achieve a larger displacement, a high-stiffness arm is used to amplify the piezoelectric displacement. To measure displacement on the order of nanometers, we use a capacitive displacement sensor because of its properties, including no contact, high accuracy, high bandwidth, and small size.

The maximum measurement span of the chosen capacitive sensor is 50 μm. The resolution can be better than 1 nm if we do not consider the system noise, such as the electrical noise the driver amplifiers. Based on the capacitive displacement sensor, we will investigate the modeling and identification of the multifield hysteretic dynamics in the flexure-guided piezoelectric platform.

Compared with high-stiffness piezoelectric platforms, the hysteretic dynamics in flexure-guided piezoelectric platforms is much more complex owing to the low-stiffness flexible joints. The hysteretic dynamics results from the multifield, which consists of the material field (i.e., hysteresis effect and creep effect [5]), the electrical field, and the mechanical field (i.e., mechanical vibration dynamics). First, the static hysteresis effect is a rate-independent nonlinearity which has global memories [6]. The classical Preisach model can be used to represent the static hysteresis [7]. Second, the creep, electrical, and mechanical vibration dynamics have different timescales [8].

The tracking accuracy can be degraded significantly by the hysteretic dynamics [9]. Even at low frequencies, the tracking error of a piezoelectric platform can be as large as 15% of the travel span [10]. Various methods have been investigated to model the hysteretic dynamics in piezoelectric actuators or platforms. For high-stiffness piezoelectric actuators or platforms, the electrical and vibration dynamics are generally on the order of kilohertz, and the creep effect and the hysteresis effect are much slower than the electrical and vibration dynamics. Thus, the hysteresis, creep, and vibration dynamics can be treated separately. For instance, Song et al. [11] used the static Preisach hysteresis to represent the piezoelectric system without modeling the electrical and vibration dynamics. We used a curve-fitting technique to represent and reduce the creep effect while identifying the static hysteresis effect [10, 12]. To apply modern robust control, Huang et al. [13] and Wu and Zou [14] used vibration dynamics to represent piezoelectric actuators.

Additionally, the hysteretic dynamics can be investigated from the mathematical perspective. For instance, Mayergoyz [15] presented a rate-dependent Preisach model with dynamic density functions, and Ge and Jouaneh [16] presented a generalized Preisach model without the limitation of the congruency property. Janaideh et al. [17] modeled the dynamic hysteresis of piezoelectric actuators using a rate-dependent Prandtl-Ishlinskii model and identified the generalized Prandtl-Ishlinskii hysteresis by assuming the density function. Jiang et al. [18] modified the Prandtl-Ishlinskii model to represent the asymmetric hysteresis of piezoelectric actuators. It is difficult to use the pure mathematical models for further development of piezoelectric platforms, because physical variables cannot be found in the pure mathematical models. Alternatively, hysteretic dynamics with some physical meaning is increasingly needed for the flexure-guided piezoelectric platform. The full modeling of the hysteretic dynamics in the flexure-guided platform has still not been solved well. Hence, in this chapter, we investigate the complete modeling of the hysteretic dynamics in flexure-guided platforms.

To fully model the hysteretic dynamics, the multifield domain is investigated for the piezoelectric platform. The multifield domain consists of the material, electrical, and mechanical fields. Conveniently, we can achieve the hysteretic dynamics in the flexure-guided platform by combining the components which represent different dynamics and field effects. Tan and Baras [19] presented a cascade connection using Preisach hysteresis and nonhysteretic dynamics. Conversely, Bree et al. [20] proposed a feedback combination of Duhem hysteresis and nonhysteretic dynamics to represent hysteretic dynamics, but it is difficult to identify parameters in the feedback combination. In this chapter, we used the cascade connection to combine the components of the effects and dynamics in the piezoelectric platform. The hysteresis effect is represented by the classical Preisach model. The nonhysteretic creep, electrical, and vibration dynamics are represented by transfer functions, such that it is easy for researchers to use the proposed hysteretic dynamics for further developments.

After modeling the hysteretic dynamics, we will identify the parameters. Multiple timescales and couplings are significant limitations, especially for the flexure-guided piezoelectric platform. The multifield hysteretic dynamics exists on multiple timescales, which increases the difficulty of model identification. The creep dynamics is a slow-timescale effect which is generally on the order of minutes. The hysteresis effect is a rate-independent nonlinearity which is independent of time [7]. For high-stiffness piezoelectric platforms, the vibration dynamics is on the order of milliseconds, but for the low-stiffness piezoelectric platform considered in this chapter, the vibration dynamics is on the order of 100 ms. The decrease of the vibration frequencies increases the couplings among the components of the hysteretic dynamics. For low-stiffness piezoelectric platforms, the couplings are not fully treated when the hysteretic dynamics is identified. Rakotondrabe et al. [21] identified the hysteresis, creep, and vibration dynamics without considering the couplings, and Juhsz et al. [22] separately identified the Maxwell resistive capacitor hysteresis and vibration dynamics without considering the couplings and the creep effect.

In this chapter, we will consider the multiple timescales and couplings between the hysteresis, creep, electrical, vibration dynamics while applying the identification.

In [8], we proposed an identification approach for high-stiffness piezoelectric actuators. In this chapter, the flexible joints are much softer, which means the vibration dynamics will have lower frequencies. As a result, the couplings among the components of the hysteretic dynamics are much more significant. Thus, we will further investigate the hysteretic dynamics by adequately considering the couplings and multiple timescales. The further identification investigation of hysteretic dynamics in the flexure-guided piezoelectric platform is also presented in this chapter. The comprehensive identification approach is designed according to the characteristics of the flexure-guided platform. A novel technique is developed to identify the electrical and vibration dynamics. Additionally, special input signals are designed to identify the static Preisach hysteresis.

This chapter is organized as follows: First, the system and the experimental setup are described in Section 4.2. Next, the complete modeling of the hysteretic dynamics in the flexure-guided piezoelectric platform is presented in Section 4.3. Multifield effects and dynamics are contained in the hysteresis model. Then, the identification strategy is developed in Section 4.4. To further validate the proposed modeling and identification approaches in this chapter, an experimental study is proposed in Section 4.5. The experimental results are discussed in Section 4.6. Finally, the conclusions are presented in Section 4.7.

4.2 Description of a Piezoelectric Smart System

4.2.1 Typical Structure of a Piezoelectric Smart System

The piezoelectric smart system, a typical microdisplacement platform, as shown in Figure 4.1, comprises a lead zirconate titanate (PZT) stack, a motion-amplifying arm, a working platform, and a capacitive sensor. The working platform moves along the flexure guide, which consists of four flexible joints. The PZT stack is driven by an input voltage. Then, its expansion is transferred to the working platform by the motion-amplifying arm. The resulting displacement is measured by the capacitive displacement sensor. To avoid friction and clearance around the working platform, flexible joints are used instead of common

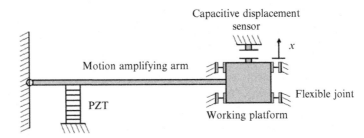

Figure 4.1 The flexure-guided piezoelectric platform.

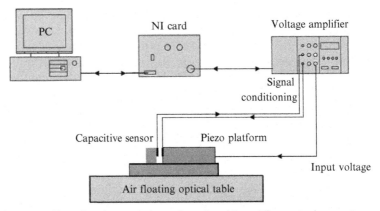

Figure 4.2 Piezoelectric smart system in a closed loop. PC, personal computer.

joints and bearings. Especially, low-stiffness flexible joints are chosen, such that low voltage can drive large displacement. This design often decreases the resonant frequency of the piezoelectric platform, which significantly increases the couplings between the hysteresis, creep, and electrical and vibration dynamics. To solve this problem, we fully model and identify the hysteretic dynamics in multifields.

The smart system in a closed loop consists of the piezoelectric platform, a capacitive displacement sensor, a voltage amplifier, a signal conditioning unit and a National Instruments (NI) control card. The piezoelectric platform and the capacitive sensor are shown in Figure 4.2. To achieve larger displacements, low-stiffness flexible joints are used instead of high-stiffness flexible joints. A motion-amplifying mechanism is also used. The maximum displacement of a typical piezoelectric platform is 50 μm. For nanometer-positioning accuracy, a noncontact capacitive displacement sensor is used. Finally, the NI control card is used to generate the voltage input and measurement output. The flexure-guided piezoelectric platform and the capacitive sensor are mounted on the air-floating table to isolate ground vibrations. The NI card is connected to a personal computer, amplifier, and signal conditioning unit. The voltage amplifier and signal conditioning unit are installed in a metallic box.

4.2.2 Working Principle of a Capacitive Displacement Sensor

A noncontact capacitive sensor is used in this chapter. It is an analog device which contains two electrodes. The electrode area and the dielectric are constant; thus, the capacitance change is a result of the distance change between the electrodes. For better stability, a high-frequency AC excitation voltage is used to drive the capacitive sensor. The components of the capacitive displacement sensor are shown in Figure 4.3. The capacitive displacement sensor consists mainly of the probe and target electrodes. The probe comprises a sensor active area, a guard area, housing, and a coaxial cable. The guard is used to prevent the electric field from spreading outside the sensing area. The measured range is set to 50 μm. The resolution is limited mainly by the electrical noise and the measured range. The relative error of the capacitive displacement can be less than 0.01% of the measured range.

The working principle of the capacitive displacement sensor is shown in Figure 4.4. The sensing electrode is driven by an alternating voltage. The guard is driven by the same voltage to avoid a voltage difference (or electric field) between the guard area and the sensor active area. It can also be seen that the sensing electric field is protected by the guard electric field, because the external electric field is stopped by the guard electric field.

For stability, a linear capacitive bridge is also used to measure the capacitance change. A sketch of the capacitive bridge is shown

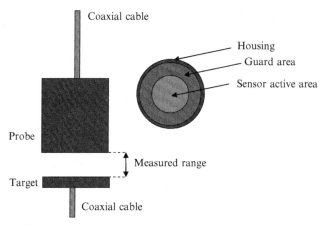

Figure 4.3 Components of the capacitive displacement sensor.

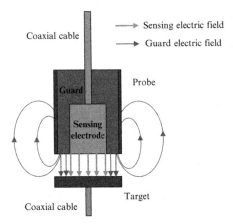

Figure 4.4 Working principle of the capacitive displacement sensor.

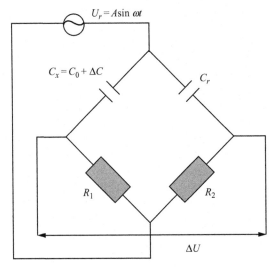

Figure 4.5 The capacitive bridge.

in Figure 4.5. R_1 and R_2 are resistors, C_r is the reference capacitor, and C_x represents the capacitive sensor. C_0 is the nominal value, ΔC is the capacitance change due to the distance change of the capacitive sensor, and ΔU is the generated voltage resulting from the distance change. $U_r = A\sin(\omega t)$ is the high-oscillation input voltage which continuously reverses the charge directions. The oscillation frequency ω is regulated near the resonant frequency of the circuit such that the linearity and sensitivity are satisfied. The balancing of the capacitive bridge is achieved by use of linear AC bridge circuits. More details can be found in Ref. [23].

4.3 Multifield Modeling of the Hysteretic Dynamics

In this section, the multifield modeling of the hysteretic dynamics in the flexure-guided platform is presented. The hysteretic dynamics model is derived from the material, electrical, and mechanical fields. The complete modeling of the hysteretic dynamics consists of the static Preisach hysteresis, the creep effect, and electrical and vibration dynamics. Additionally, the variables in the complete hysteretic dynamics have physical meanings, which is more beneficial for further developments for researchers and engineers.

4.3.1 Multifield Modeling of the Hysteretic Dynamics

The multifield contains the material, electrical, and mechanical fields, in which the hysteresis effect, the creep effect, and the electrical and vibration dynamics are derived, respectively. First, we present the hysteresis effect due to the PZT stack is using the classical Preisach model. Next, we derive the electrical dynamics in the electric field. Then, we derive the mechanical vibration dynamics using stiffness and damping parameters. Additionally, we present the creep effect using a linear transfer function. Finally, the characteristics of the hysteretic dynamics are proposed. The hysteresis effect and electrical dynamics are shown in Figure 4.6.

H represents the hysteresis effect, R and C represent the resistance and capacitance of the PZT stack, respectively, T_{em} represents the electromechanical transformer ratio of the PZT material, i is the conductor current, u is the input voltage of the PZT stack, u_h is the voltage drop due to the hysteresis effect, and u_p is the effective voltage for the PZT stack.

First, the hysteresis effect is presented. The hysteresis between the input voltage and the effective PZT voltage can be represented as the following Preisach model [7]:

$$u_p = \iint_{S^+} \mu(\alpha, \beta) \gamma_{\alpha\beta}[u(t)] \, d\alpha \, d\beta, \tag{4.1}$$

where u_p is the hysteresis output, and $\mu(\alpha, \beta)$ and $\gamma_{\alpha\beta}$ are the density function and hysteron output of point α, β on the Preisach plane, respectively. The Preisach model is rate independent (i.e., it is a static model). Figure 4.7 shows the Preisach plane, where S^+

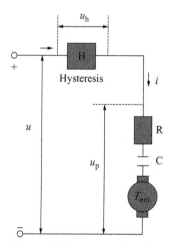

Figure 4.6 Hysteresis effect, electrical dynamics, and inverse piezoelectric effect (u_h is the voltage drop due to the hysteresis effect, and u_p is the effective voltage of R and C).

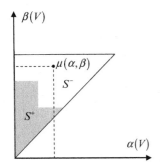

Figure 4.7 Preisach plane: S^+ represents the actuated area with $\gamma_{\alpha\beta}$ of 1, S^- represents the unactuated area with $\gamma_{\alpha\beta}$ of 0, and $\mu(\alpha, \beta)$ represents the density function at the point (α, β).

represents the actuated area with $\gamma_{\alpha\beta}$ of 1 and S^- represents the unactuated area with $\gamma_{\alpha\beta}$ of 0.

In the electric field of the PZT stack, the input voltage is represented as

$$u = u_h + u_p, \tag{4.2}$$

where u is the input voltage of the PZT stack, u_h is the voltage drop due to the hysteresis effect, and u_p is the effective voltage of R and C.

The voltage drop u_p is represented by

$$u_p = iR + u_c, \tag{4.3}$$

where R is the resistance, i is the current, and u_c is the voltage of the equivalent capacitor of the PZT stack. u_c can be represented by

$$u_c = QC, \tag{4.4}$$

where Q is the charge and C is the equivalent capacitance of the PZT stack. Additionally, the conduction current i is represented by

$$i = \frac{dQ}{dt}. \tag{4.5}$$

By combining Equations (4.2)-(4.4), we can write the electrical dynamics as

$$\frac{Q(s)}{u_p(s)} = \frac{C}{(1 + \tau s)}, \tag{4.6}$$

where s is the Laplace operator $\tau = RC$.

The force F due to the inverse piezoelectric effect of the PZT stack is written as

$$F = T_{em}Q, \tag{4.7}$$

where T_{em} represents the electromechanical transformer ratio resulting from the inverse piezoelectric effect. The mechanical vibration dynamics of the flexure-guided piezoelectric platform is shown in Figure 4.8. K_p and C_p represent the stiffness and damping of the PZT stack. K_f and C_f represent the stiffness and damping of the flexure guide, which comprises four flexible joints. x represents the displacement of the working platform. θ represents the tilt angle due to the piezoelectric displacement. L represents the

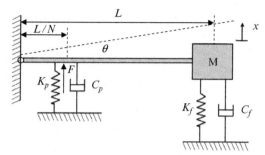

Figure 4.8 Mechanical vibration dynamics of the piezoelectric platform
(L represents the length of the motion-amplifying arm, N represents the amplifying
value of the motion-amplifying arm, K_p and C_p represent the stiffness and
damping of the PZT stack, K_f and C_f represent the stiffness and damping of the
flexure guide, which comprises four flexible joints, and x represents the
displacement of the working platform).

length of the motion-amplifying arm. N represents the amplifying
value of the motion-amplifying arm.

In the flexure-guided piezoelectric platform, only microdis-
placements are provided. The maximum displacement is 50 μm.
Compared with $L = 10$ cm, $\tan\theta = x/L < 0.0005$. Thus, $\theta \doteq \tan\theta$,
$x = L\theta$. According to Newton's second law, the dynamics of M can
be written as

$$J\ddot{\theta} + K_f x L + C_f \dot{x} L + K_p x_p \frac{L}{N} + C_p \dot{x}_p \frac{L}{N} = F\frac{L}{N}, \qquad (4.8)$$

where $J = ML^2$, $x_p = x/N$, K_f and C_f are the stiffness and damping
of the piezoelectric platform, and K_p and C_p are the stiffness and
damping of the PZT stack.

Equation (4.8) can be written as

$$M\ddot{x} + \left(C_f + \frac{C_p}{N^2}\right)\dot{x} + \left(K_f + \frac{K_p}{N^2}\right)x = \frac{F}{N}. \qquad (4.9)$$

Then,

$$\frac{X(s)}{Q(s)} = k_v \frac{\omega_n^2}{s^2 + 2\xi_n \omega_n s + \omega_n^2}, \qquad (4.10)$$

where $\omega_n = \sqrt{\left(C_f + \frac{C_p}{N^2}\right)/M}$, $2\xi_n\omega_n = \left(K_f + \frac{K_p}{N^2}\right)/M$,
$k_v = T_{em}/\left(K_f + K_p/N^2\right)$.

By combining Equations (4.6) and (4.10), we can represent the electrical and resonant dynamics as

$$G_{ev} = k_{ev} \frac{1}{1 + \tau s} \frac{\omega_n^2}{s^2 + 2\xi_n \omega_n s + \omega_n^2}, \qquad (4.11)$$

where $k_{ev} = k_v C$. Further, high-frequency vibration modes can be added to the derived electrical and vibration dynamics,

$$G_{ev} = k_{ev} \frac{1}{1 + \tau s} \frac{\omega_n^2}{s^2 + 2\xi_n \omega_n s + \omega_n^2} \prod_j \frac{s^2 + 2\bar{\xi}_j \bar{\omega}_j s + \bar{\omega}_j^2}{s^2 + 2\xi_j \omega_j s + \omega_j^2}, \qquad (4.12)$$

where ξ_j and ω_j represent the damping and resonant frequency of the poles of mode j, and $\bar{\xi}_j$ and $\bar{\omega}_j$ represent the damping and resonant frequency of the zeros of mode j. Finally, there exists the creep effect (also named drift) in PZT material. The creep effect can be represented by [24]

$$G_c(s) = k_c \prod_l \frac{s + z_{cl}}{s + p_{cl}}, \qquad (4.13)$$

where k_c is the creep gain when s goes to infinity (i.e., k_c represents the creep gain at infinite frequencies), l is the creep order, and p_{cl} and z_{cl} are the poles and zeros of the creep dynamics. The relationship of the multifield hysteretic dynamics is shown in Figure 4.9. The cascade connection is used to represent the relationship among the components of the hysteretic dynamics. The hysteresis effect and the creep effect are included in the material field, the electrical dynamics is included in the electric field, and the vibration dynamics is included in the mechanical field. Figure 4.10 shows the responses of multi-field hysteretic dynamics under square wave inputs. It can be seen that the creep exhibits slower time scale compared with the electric and mechanical dynamics.

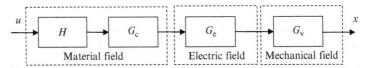

Figure 4.9 Multifield dynamics of the flexure-guided piezoelectric platform (H, G_c, G_e, and G_v represent the hysteresis effect, creep effect, electrical dynamics, and vibration dynamics, respectively).

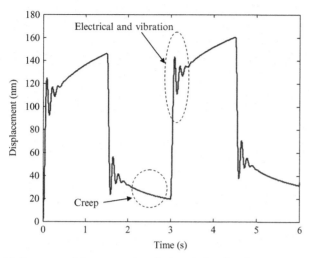

Figure 4.10 Response of the creep, electrical, and vibration dynamics.

4.4 Identification Strategy Design

The comprehensive identification strategy for the hysteretic dynamics in the piezoelectric platform is proposed in this section. According to the proposed properties and characteristics, the components of the hysteretic dynamics are identified one by one, because the components have different timescales. First, pre-execution of the nonhysteretic dynamics is presented. Next, the identification approach for the nonhysteretic creep, electrical, and mechanical dynamics is proposed. Then, the identification strategy for the Preisach hysteresis is presented.

4.4.1 Pre-execution of the Creep, Electrical, and Vibration Dynamics

The pre-execution of the creep, electrical, and vibration dynamics is proposed to satisfy the linear-in-parameter condition and reduce the coupling between the hysteresis, creep, electrical dynamics, and vibration models. The parameters k_c and k_{ev} in the nonhysteretic dynamics are absorbed by the Preisach hysteresis. Otherwise, the term $k_c k_{ev} \mu(\alpha, \beta)$ is not linear in parameter (i.e., k_c and k_{ev} cannot be separated in the identification). Thus, without loss of generality, $k_c k_{ev} \mu(\alpha, \beta)$ is regarded as a new density function when the creep, electrical, and vibration dynamics are identified. This treatment has the following merits: (1) the linear-in-parameter condition is satisfied; (2) the couplings are reduced. It is easy to implement the identification one by one with different inputs.

4.4.2 Identification of the Creep, Electrical, and Vibration Dynamics

The timescale of the creep effect is the largest among the components. Thus, the creep effect is identified first, in which the error due to the electrical and vibration dynamics is reduced by use of a slow sampling rate. Conversely, if the electrical and vibration dynamics are identified first, it is difficult to separate the creep effect from the displacement output, and the identification accuracy will decrease. The identification of the creep dynamics is shown in Figure 4.11, where $\hat{v}_h$ is the estimation of the Preisach hysteresis output v_h, and k_H represents an uncertainty gain resulting from Preisach hysteresis with a square-wave input.

We use a square input with a period on the order of minutes for the flexure-guided piezoelectric platform. Under the square input, the Preisach hysteresis behaves as the factor k_H, which is an uncertain value with respect to square inputs with different amplitudes. In the pre-execution section, k_H can be absorbed by the Preisach hysteresis. The input voltage u can be regarded as the creep input. We can use the piezoelectric displacement y as the creep output v by using slow sampling rates, by which the response of the electrical and vibration dynamics can be reduced. Thus, we can identify the creep dynamics using u and y.

Next, the electrical and vibration dynamics can be identified after the creep identification. The identification of the electrical and vibration dynamics is shown in Figure 4.12, where $\hat{G}_c$ represents the identification of G_c, and $\hat{v}_c$ represents the estimation of the creep effect output v_c. To identify the electrical and vibration dynamics (i.e., G_e and G_v), $\hat{v}_c$ is regarded as their estimated input. The square input at higher frequencies is used, and the sampling rate is set as fast as possible to capture the responses of the electrical and vibration dynamics.

The identification is implemented in the discrete domain. To obtain the model structure in discrete form, the transfer function $G_{ev}(s)$ is transformed to its discrete form by the following Tustin transformation:

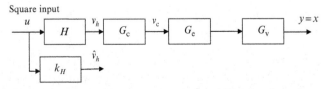

Figure 4.11 Identification of the creep, electrical, and vibration dynamics (v_h represents the hysteresis output, v_c represents the creep output, and $\hat{v}_h$ represents the estimation of v_h).

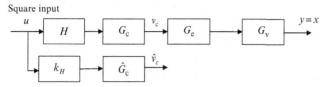

Figure 4.12 Identification of the electrical and vibration dynamics ($\hat{v}_c$ represents the estimation of v_c).

$$s = \frac{2}{T}\frac{1 - z^{-1}}{1 + z^{-1}}, \tag{4.14}$$

where T is the sampling interval and s is the discrete operator.

Let $G_{ev} = X(s)/V(s)$. Then, the discrete form of the electrical and vibration dynamics is written as

$$\frac{x(z)}{v(z)} = \frac{\left(1 + z^{-1}\right)^3 \left(b_{m+1}z^{-m+2} + b_m z^{-m} + \cdots + b_1 z^{-1} + b_0\right)}{a_{m+1}z^{-m-1} + a_m z^{-m} + \cdots + a_1 z^{-1} + a_0}. \tag{4.15}$$

The discrete transfer function G_{ev} is reformulated as

$$\frac{x(z)}{v(z)\left(1 + z^{-1}\right)^3} = \frac{b_{m+1}z^{-m+2} + b_m z^{-m} + \cdots + b_1 z^{-1} + b_0}{a_{m+1}z^{-m-1} + a_m z^{-m} + \cdots + a_1 z^{-1} + a_0}, \tag{4.16}$$

where m is the order of the vibration dynamics, and $a_{m+1}, a_m, \ldots, a_0$ are the coefficients of $z^{-m-1}, z^{-m}, \ldots, z^{-1}$, respectively.

Let $w(z) = \left(1 + z^{-1}\right)^3 v(z)$. Then, $w(k)$ at time instant k is computed by

$$w(k) = v(k - 3) + 3v(k - 2) + 3v(k - 1) + v(k). \tag{4.17}$$

Then, the discrete transfer function is identified according to (4.16). The final identification result for the electrical and vibration dynamics is computed by

$$G_{ev} = \frac{x(z)}{w(z)}\left(1 + z^{-1}\right)^3. \tag{4.18}$$

4.4.3 Identification of the Preisach Hysteresis

The Preisach hysteresis is identified with use of sinusoidal inputs with varying amplitudes. According to our investigation of the Preisach hysteresis identification, inputs with adequate amplitudes can satisfy the persistent-excitation (PE) condition. The identification of the Preisach hysteresis is shown in Figure 4.13, where $\hat{G}_{cev}$ represents the estimation of the creep, electrical, and

Sinusoidal input

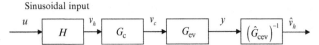

Figure 4.13 Identification of the Preisach hysteresis ($\hat{G}_{\text{cev}}$ represents the estimation of G_{cev}, where $G_{\text{cev}} = G_c G_e G_v$).

vibration dynamics, and $\hat{v}_{\text{h}}$ represents the output of $\hat{G}_{\text{cev}}$. The effects of the creep, electrical, and vibration dynamics on the piezoelectric displacement are compensated by their inversion $\left(\hat{G}_{\text{cev}}\right)^{-1}$. As a result, the Preisach output v_{h} is estimated by $\hat{v}_{\text{h}}$.

To identify the Preisach hysteresis, the discrete form is derived. The hysteresis output at time k can be written as the following discrete form:

$$v_h(k) = A_k X, \tag{4.19}$$

where $A_k = [\gamma_{11}(k), \gamma_{21}(k), \gamma_{22}(k), \ldots, \gamma_{NN}(k)]$ and $X = [\mu_{11} s_{11}, \mu_{21} s_{21}, \mu_{22} s_{22}, \ldots, \mu_{NN} s_{NN}]$, where s_{NN} is the grid area on the Preisach plane.

Then, the linear equation (4.19) can be given by

$$AX = Y, \tag{4.20}$$

where $A = \begin{bmatrix} A_1^{\text{T}} & \cdots & A_q^{\text{T}} \end{bmatrix}$ and $Y = \begin{bmatrix} v_{\text{h}}(1) & \cdots & v_{\text{h}}(q) \end{bmatrix}$, where q is the sampling number. The estimation of X can be obtained with the pseudoinverse

$$X = A^+ Y, \tag{4.21}$$

where A^+ is the pseudoinverse of A; more details can be found in Ref. [12].

We design special inputs and sampling rules to identify the Preisach hysteresis. Preisach hysteresis is a static and rate-independent nonlinearity. Adequate frequency is not beneficial to satisfy the PE condition of the Preisach hysteresis identification. Instead, adequate amplitudes are sufficient to satisfy the special PE condition. In this chapter, sinusoidal inputs with varying amplitudes in (4.22) are used instead of inputs with varying frequencies [8].

$$u(t) = \frac{P(t)}{2}(1 - \cos(\omega_{\text{r}} t)), \tag{4.22}$$

where ω_{r} is the frequency of each sinusoidal signal and $P(t)$ is the stair amplitude of each harmonic signal, and is given by $P(t) = \text{fix}\{1 + [t - \text{fix}(t/T)]T\omega_{\text{r}}\} \cdot \delta$, where T is the whole period of $P(t), \delta$

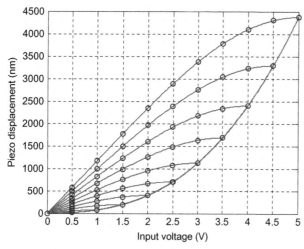

Figure 4.14 Sampling in hysteresis loops (circles represent samples for Preisach hysteresis identification and solid lines represent the hysteresis loops).

is the discretization level of the maximum input voltage $u(t)$, and fix represents a round function toward zero (i.e., fix(x) rounds the elements of x to the nearest integers toward zero).

Furthermore, a special sampling with respect to the inputs in (4.22) is given by the following equation:

$$\begin{cases} t_{i,j} = \arccos\left(1 - \frac{2j\delta}{P(t)}\right)/\omega_r + t_{i,1}, \\ t_{i,j} = \left[2\pi - \arccos(1 - \frac{2j\delta}{P(t)})\right]/\omega_r + t_{i,1}, \end{cases} \quad (4.23)$$

where $j = 1, 2, \ldots, P/\delta$, and $t_{i,1}$ is the starting time of the ith harmonic signal.

Figure 4.14 shows the sampling rules of the hysteresis loops. The sinusoidal inputs change their amplitudes after each period, resulting in incremental hysteresis loops. The samplings of the input voltages are uniform with voltage, but are is not uniform with time. The displacement samples (i.e., samples for Preisach hysteresis identification) corresponding to the input voltages are computed with (4.23).

4.5 Experimental Studies of the Proposed Modeling and Identification

4.5.1 Experimental Setup

Figure 4.15 illustrates the experimental setup.

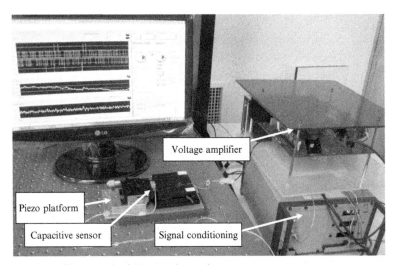

Figure 4.15 Photograph of the experimental setup.

4.5.2 Identification Result for the Creep, Electrical, and Vibration Dynamics

The proposed comprehensive identification of the hysteretic dynamics in the flexure-guided piezoelectric platform is implemented in experiments. We use the autoregressive moving average model with external input method to compute the creep, electrical, and vibration dynamics [25]. The sampling rate for the creep dynamics identification is 20 Hz. The sampling rate for the electrical and vibration dynamics identification is 500 Hz.

Using the identification strategy in Section 4.4, we obtain the creep dynamics as

$$G_{\mathrm{c}} = \frac{(s+6.73)\,(s+1.47)\,(s+0.235)}{(s+6.62)\,(s+1.42)\,(s+0.225)}. \tag{4.24}$$

The electrical and vibration dynamics are identified as

$$G_{\mathrm{ev}} = \frac{40.5}{(s+35.22)} \frac{3510}{(s^2+38.84s+3510)} \frac{(s^2-940.5s+3.2\times10^5)}{(s^2+460.2s+3.388\times10^5)}, \tag{4.25}$$

where RC is $1/35.22 = 0.0284$, and the first resonant frequency and damping ratio of the vibration dynamics are 9.43 Hz and 0.33, respectively. To validate the identification accuracy, the estimated and measured displacements of the piezoelectric platform under the square input are shown in Figure 4.16.

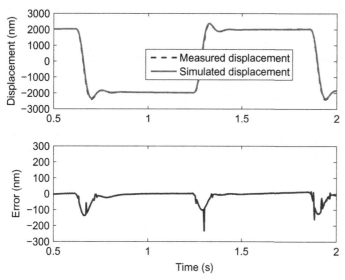

Figure 4.16 Estimated and measured displacements of the piezoelectric platform (the dtrend function is used to treat the displacement).

To represent the error more clearly, the toot mean square RMS relative error in (4.26) is used [14]:

$$
e_{\text{RMS}} = \left(\frac{\sqrt{\frac{1}{n} \sum_{i=1}^{n} (\hat{y}(i) - y(i))^2}}{\sqrt{\frac{1}{n} \sum_{i=1}^{n} y(i)^2}} \right) \times 100\%, \qquad (4.26)
$$

where n is the sampling number, $\hat{y}$ is the estimated displacement, and y is the output displacement.

The RMS relative error of the square inputs is as small as 3.2%. It can be seen that the error between the measured and simulated displacements is acceptable.

4.5.3 Identification Result for the Preisach Hysteresis

To identify the Preisach hysteresis, a sinusoidal input with varying amplitude is used. The maximum input voltage u_{max} is 50 V. The discretization level δ is set to 1 V. The frequency ω_r of each sinusoidal signal is set to 0.2 Hz. The whole period T of $P(t)$ is set to 250 s. Then, the input voltage is given by

$$
u(t) = \text{fix} \left\{ (1 + 100\pi \left[t - \text{fix} \left(\frac{t}{250} \right) \right] \right\} \frac{1 - \cos(0.4\pi t)}{2}. \qquad (4.27)
$$

The sampling instants are given by

$$
\begin{cases}
t_{i,j} = \arccos\left(1 - \dfrac{2j}{\text{fix}\{(1+100\pi[t-\text{fix}(\frac{t}{250})]\}}\right)\Big/ (0.4\pi) + t_{i,1}, \\[2ex]
t_{i,j} = \left(2\pi - \arccos(1 - \dfrac{2j}{\text{fix}\{1+100\pi[t-\text{fix}(\frac{t}{250})]\}}\right)\Big/ (0.4\pi) + t_{i,1},
\end{cases}
\tag{4.28}
$$

where $j = 1, 2, \ldots, \text{fix}\{1 + 100\pi[t - \text{fix}(t/250)]\}$, and $t_{i,1}$ is the starting time of the ith harmonic signal. Furthermore, the effects of the creep, electrical, and vibration dynamics on the displacement response of the piezoelectric platform are compensated by their inversions, as shown in Figure 4.13.

To identify the Preisach hysteresis, the drift due to the creep, electrical, and vibration dynamics should be compensated in the output displacements. Moreover, Figure 4.17 shows the incremental hysteresis loops with drift compensation. It can be seen that the drift is efficiently reduced by the inversion $\left(\hat{G}_{\text{cev}}\right)^{-1}$. With the inversion-based drift compensation, the RMS error of the displacement samples corresponding to zero voltage is 4.7 nm. Then, the Preisach hysteresis is identified with (4.21). Figure 4.18 shows the identified density function $\mu(\alpha, \beta)$ of the Preisach hysteresis, where $\mu(\alpha, \beta)$ is a distributed function of the voltage variables α and β in the Preisach plane.

To validate the identified creep, electrical, and vibration models, the measured and simulated displacements are compared, as shown in Figure 4.19, in which $\hat{H}$, $\hat{G}_{\text{c}}$, $\hat{G}_{\text{e}}$, and $\hat{G}_{\text{v}}$ represent the estimation of H, G_{c}, G_{e}, and G_{v}, respectively, and $\hat{v}_{\text{h}}$, $\hat{v}_{\text{c}}$, and $\hat{y}$

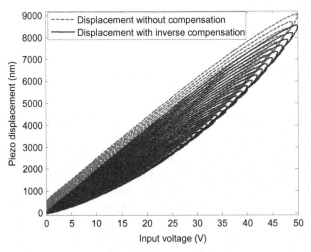

Figure 4.17 Hysteresis loops with drift compensation.

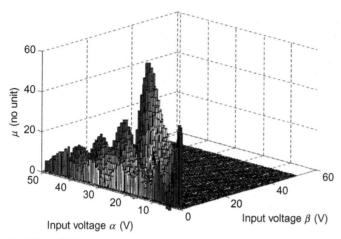

Figure 4.18 Identified density function $\mu(\alpha, \beta)$ of the Preisach hysteresis.

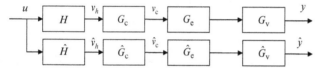

Figure 4.19 The identification validation ($\hat{H}$, $\hat{G}_c$, $\hat{G}_e$, and $\hat{G}_v$ represent the estimation of H, G_c, G_e, and G_v, respectively, and $\hat{v}_h$, $\hat{v}_c$, and $\hat{y}$ represent v_h, v_c, and y, respectively).

represent v_h, v_c, and y, respectively. Figure 4.20 shows the measured and simulated displacements under the input with varying amplitudes. The RMS relative error is 5.5%. Further, the nonperiodic input is used to validate the effectiveness of the identified hysteretic dynamics. The experimental results indicate that the identified hysteretic dynamics is effective to model the flexure-guided piezoelectric platform.

4.5.4 Discussion

Compared with high-stiffness piezoelectric platforms, flexure-guided piezoelectric platforms exhibit more complex hysteretic dynamics, because the couplings between the hysteresis, creep, electrical, and vibration dynamics are more significant in the flexure-guided piezoelectric platform. To satisfy the requirement of the accurate modeling and identification in the process of platform development, we have fully modeled the hysteretic dynamics by including the hysteresis and creep, electrical dynamics, and vibration dynamics in the material, electrical, and mechanical

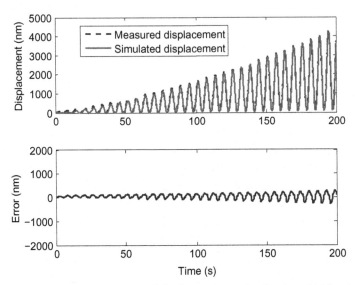

Figure 4.20 Estimated and measured displacements under the sinusoidal input.

fields, respectively. The proposed modeling method was validated in the experimental study.

The serious couplings between the hysteresis, creep, electrical dynamics, and vibration dynamics significantly increase the difficulty of further developments based on the accurate hysteretic dynamics. The proposed complete modeling and comprehensive identification is beneficial for further developments of the flexure-guided piezoelectric platform.

The experimental results reveal that the proposed modeling and identification is effective. Some parameters of the proposed hysteretic dynamics have clear physical meanings, such as the time constant of the electrical dynamics, and the resonant frequency and damping ratio of the vibration dynamics. Thus, it is much easier for engineers and researchers to design or develop piezoelectric-based platforms by using the proposed hysteretic dynamics. For example, the hysteresis effect in the PZT stack can be avoided in theory by use of charge control instead of the voltage control in this chapter, as shown in (4.7), but charge amplifiers should be developed first. If the dynamic response of a piezoelectric platform is designed, both the time constant of the electrical dynamics and the stiffness and damping can be regulated with the proposed hysteretic dynamics. Furthermore, various model-based control approaches can be designed with the proposed hysteretic dynamics.

4.6 Complete Modeling of Hysteretic Dynamics in Piezoelectric Smart Systems with High Stiffness

In the previous section, the multifield modeling of hysteretic dynamics was presented for soft piezoelectric smart systems with low resonant frequencies. In this section, we investigate the multifield modeling of piezoelectric smart systems with high stiffness. In these hard smart systems, the mechanical dynamics is on the order of kilohertz, and the circuit dynamics is easily on the order of kilohertz, such that the resonant dynamics of the driving circuit in smart systems is suggested to be mechanical vibration dynamics.

The electrical dynamics contains the capacitor dynamics due to piezoelectric materials and the circuit resonant dynamics. The augmented circuit resonant dynamics is written as

$$G_{\mathrm{cr}} = \frac{\omega_{\mathrm{nc}}^2}{s^2 + 2\xi_{\mathrm{nc}}\omega s + \omega_{\mathrm{nc}}^2}, \qquad (4.29)$$

where ξ_{nc} and ω_{nc} denote the damping ratio and resonant frequency of the circuit dynamics.

The nominal mechanical vibration and electrical dynamics are synthesized as

$$G_{\mathrm{e}} = k_{\mathrm{c}} \frac{1}{RCs + 1} \frac{\omega_{\mathrm{n}}^2}{s^2 + 2\xi_{\mathrm{n}}\omega s + \omega_{\mathrm{n}}^2} \frac{\omega_{\mathrm{n}}^2}{s^2 + 2\xi_{\mathrm{nc}}\omega s + \omega_{\mathrm{nc}}^2}, \qquad (4.30)$$

where k_{c} is the DC gain.

Then, by synthesizing the static Preisach hysteresis and creep effect, we achieve the complete modeling of piezoelectric smart systems with high stiffness. We can use the same identification approach as used for soft piezoelectric smart systems.

Finally, we illustrate the typical mechanical vibration, electrical dynamics, and creep effect of piezoelectric smart systems with high stiffness. Figure 4.21 shows the Bode diagram of the typical mechanical and electrical dynamics in smart systems with high stiffness. Figure 4.22 shows the Bode diagram of the typical creep effect in smart systems with high stiffness. It can be seen that the mechanical vibration and electrical dynamics are much faster than the creep effect.

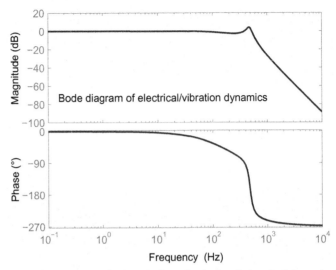

Figure 4.21 Bode diagram of the typical mechanical and electrical dynamics in smart systems with high stiffness.

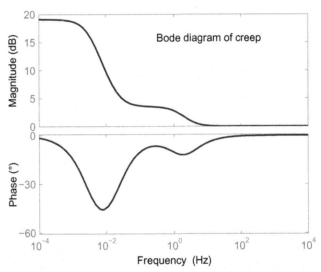

Figure 4.22 Bode diagram of the typical creep effect in smart systems with high stiffness.

4.7 Conclusion

The modeling, identification, and control of the hysteretic dynamics in piezoelectric smart systems have still not been solved well. To overcome this problem, the complete modeling and comprehensive identification by means of capacitive displacements was developed in this chapter. To begin with, the experimental setup of soft piezoelectric smart systems was presented. Then, by combining the hysteresis effect, the creep effect, and electrical and vibration dynamics, we derived the multifield hysteretic dynamics of the piezoelectric platform. After the modeling, we presented the comprehensive identification of the hysteretic dynamics. To validate the effectiveness of the proposed modeling and identification approaches, an experimental study was provided. The experimental results demonstrate the effectiveness of the proposed modeling and identification approaches. The complete modeling method for the multifield dynamics in soft smart systems was extended to hard piezoelectric smart systems.

The proposed modeling, identification, and control approaches are beneficial for further developments using the accurate hysteretic dynamics. For instance, to achieve high-accuracy tracking and positioning, model-based modern controllers can be designed with the accurate hysteretic dynamics.

References

[1] S.O. Moheimani, Accurate and fast nanopositioning with piezoelectric tube scanners: emerging trends and future challenges, Rev. Sci. Instr. 79 (2008) 071101.

[2] L. Liu, Y.G. Bai, D.L. Zhang, Z.G. Wu, Ultra-precision measurement and control of angle motion in piezo-based platforms using strain gauge sensors and a robust composite controller, Sensors 13 (7) (2013) 9070-9084.

[3] A. Putra, S. Huang, K.K. Tan, Design, modeling, and control of piezoelectric actuators for intracytoplasmic sperm injection, IEEE Trans. Control Syst. Technol. 15 (5) (2007) 879-890.

[4] H. Hemmati, A. Biswas, I. Djordjevic, Deep-space optical communications: future perspectives and applications, Proc. IEEE 99 (2011) 2020-2039.

[5] H. Xie, M. Rakotondrabe, S. Rgnier, Characterizing piezoscanner hysteresis and creep using optical levers and a reference nanopositioning stage, Rev. Sci. Instr. 80 (2009) 046102.

[6] M.A. Krasnosel'skii, A.V. Pokrovskii, Systems with Hysteresis, Springer-Verlag, New York, 1989 (translated from Russian by Marek Niezgodka).

[7] I.D. Mayergoyz, Mathematical Models of Hysteresis, Springer-Verlag, Berlin, 2003.

[8] L. Liu, K.K. Tan, C.S. Teo, S.L. Chen, T.H. Lee, Development of an approach toward comprehensive identification of hysteretic dynamics in piezoelectric actuators, IEEE Trans. Control Syst. Technol. 21 (5) (2013) 1834-1845.

[9] X. Zhang, Y. Tan, M. Su, Modeling of hysteresis in piezoelectric actuators using neuralnetworks, Mech. Syst. Signal Process. 23 (2009) 2699-2711.

[10] L. Liu, K.K. Tan, T.H. Lee, SVD-based Preisach hysteresis identification and composite control of piezo actuators, ISA Trans. 51 (3) (2012) 430-438.

[11] G. Song, J. Zhao, X. Zhou, J. Garcia, Tracking control of a piezoceramic actuator with hysteresis compensation using inverse Preisach model, IEEE/ASME Trans. Mechatron. 10 (2005) 198-209.

[12] L. Liu, K.K. Tan, S.L. Chen, C.S. Teo, T.H. Lee, Discrete composite control of piezoelectric actuators for high speed precision scanning, IEEE Trans. Indust. Informat. 9 (3) (2013) 859-868.

[13] D. Huang, J. Xu, T. Huynh, High performance tracking of piezoelectric positioning stage using current-cycle iterative learning control with gain scheduling, IEEE Trans. Ind. Electron. 61 (2) (2014) 1085-1098.

[14] Y. Wu, Q. Zou, Robust inversion-based 2-Dof control design for output tracking: piezoelectric-actuator example, IEEE Trans. Control Syst. Technol. 17 (5) (2009) 1069-1082.

[15] I.D. Mayergoyz, Dynamic Preisach models of hysteresis, IEEE Trans. Magn. 24 (1988) 2925-2927.

[16] P. Ge, M. Jouaneh, Generalized Preisach model for hysteresis nonlinearity of piezoceramic actuators, Precis. Eng. 20 (2) (1997) 99-111.

[17] M.A. Janaideh, S. Rakheja, C.Y. Su, An analytical generalized Prandtl-Ishlinskii model inversion for hysteresis compensation in micropositioning control, IEEE/ASME Trans. Mechatron. 16 (4) (2011) 734-744.

[18] H. Jiang, H. Ji, J. Qiu, Y. Chen, A modified Prandtl-Ishlinskii model for modeling asymmetric hysteresis of piezoelectric actuators, IEEE Trans. Ultrason. Ferroelect. Frequen. Control 57 (2010) 1200-1210.

[19] X. Tan, J.S. Baras, Adaptive identification and control of hysteresis in smart materials, IEEE Trans. Autom. Control 50 (6) (2005) 827-839.

[20] P. Bree, C. Lierop, P. Bosch, Feedforward initialization of hysteretic systems, in: 49Th IEEE Conference on Decision and Control, Atlanta, USA, 2010, pp. 3505-3510.

[21] M. Rakotondrabe, C. Clvy, P. Lutz, Complete open loop control of hysteretic, creeped, and oscillating piezoelectric cantilevers, IEEE Trans. Automat. Sci. Eng. 7 (3) (2010) 440-450.

[22] L. Juhasz, J. Maas, B. Borovac, Parameter identification and hysteresis compensation of embedded piezoelectric stack actuators, Mechatronics 21 (1) (2011) 329-338.

[23] P. Holmberg, Automatic balancing of linear AC bridge circuits for capacitive sensor elements, IEEE Trans. Instr. Measur. 44 (1995) 803-805.

[24] L.E. Malvern, Introduction to the Mechanics of a Continuous Medium, Prentice Hall, Englewood Cliffs, NJ, 1969.

[25] L. Ljung, System Identification: Theory for the User, Prentice Hall, Upper Saddle River, NJ, 1999.

5

CONTROL APPROACHES FOR SYSTEMS WITH HYSTERESIS

CHAPTER OUTLINE

Abstract

This chapter investigates control approaches for smart systems with hysteresis. Proportional-integral-derivative (PID) tuning control, robust control, inversion-based feedforward control, and multirate-based composite control are investigated. The PID tuning method is modified according to the hysteresis

Modeling and Precision Control of Systems with Hysteresis
http://dx.doi.org/10.1016/B978-0-12-803528-3.00005-7

effect. Inversion-based feedforward control is developed with the composite representation of hysteresis. Then, composite control involving PID feedback control and inversion-based feedforward control is presented. Moreover, robust control is also investigated in smart systems. Finally, multirate composite robust control is designed to achieve both high bandwidth and precision tracking by use of a dSPACE 1104 board.

Keywords: Control approach, Systems with hysteresis, PID tuning, Robust control

5.1 Introduction

Various control approaches have been investigated in smart systems with hysteresis. Four types of control approaches are mainly proposed as follows.

Proportional-integral-derivative (PID)-based classical control. Most smart systems have open-loop responses of high bandwidth, such as the open-loop response of typical piezoelectric actuators on the order of kilohertz. At low frequencies, the smart system has a DC gain with an input uncertainty, because the responses of the mechanical vibration and electrical dynamics are weak, and the static hysteresis contributes 5-15% of the output displacement. Creep is a slow effect and it is easily compensated by integral feedback. PID control is thus suitable for application in smart systems. In the literature, PID, proportional-integral (PI), and integral control are widely applied in smart system-based precision instruments. If engineers are not familiar with hysteresis theory and control theory, it is suggested they use PI or integral control of smart systems.

Hysteresis-based feedforward control. To further explore the performance of smart systems, PID control is not adequate. PID control approaches can provide precision motion at low frequencies. As the tracking frequency increases, the performance of PID control is degraded significantly. In the experiment, the bandwidth of PID feedback control is typically less than one third of the resonant frequency to avoid possible oscillations. Thus, hysteresis-based feedforward control approaches are investigated first to enhance the tracking performance. Static hysteresis, such as Preisach hysteresis and Prandtl-Ishlinskii hysteresis, are first used to further improve the PID feedback performance. Then, dynamic hysteresis-based feedforward control approaches are investigated to improve the tracking performance at high frequencies. Finally, composite-hysteretic-dynamics-based feedforward control can enhance the tracking bandwidth as far as possible. The hysteresis-based feedforward control approaches are complex. Engineers are required to be familiar with hysteresis theory.

Nonhysteretic dynamics-based modern control. Linear dynamics of smart systems, mainly mechanical vibrations, are frequently used to design modern controllers. Sometimes, the friction model using an ordinary differential equation, such as the Bouc-Wen model, is used as a feedback element to represent the hysteresis effect. The modeling approaches for smart systems are similar to those for other common systems; thus, various control approaches can be developed, such as linear dynamics-based robust control, nonlinear control, and adaptive control.

Nonhysteretic dynamics-based intelligent control. Modeling of smart systems is generally time invariant. The responses of smart systems are very repeatable under the same input voltage, such that intelligent control approaches are suitable to track a periodic signal. Repetitive control is suggested. Neural network-based control is another suitable choice.

In this chapter, we modify the PID tuning using the hysteresis effect. Inversion-based feedforward control is also developed. Robust control is designed with weighing functions and the loop-shaping technique. Finally, composite robust control is presented. Composite representation-based inversion-based feedforward is used to compensate the hysteretic dynamics and improve the tracking velocity. Robust H_∞ control is designed to compensate disturbances and provide precision tracking.

5.2 PID Control Tuning

5.2.1 Ziegler-Nichols Tuning Control

First, Ziegler-Nichols tuning control is described. A relay is used to test the ultimate gain and ultimate period. The classical PID tuning method using a relay can be found in various textbooks [1]. Under a relay with amplitude of d, the ultimate gain of the PID tuning can be written as [2]

$$k_\mathrm{u} = \frac{4d}{\pi a},\tag{5.1}$$

where $2d$ is the relay amplitude (peak to peak), and a is the output amplitude (peak to peak).

The ultimate period p_u is measured when the relay feedback becomes steady oscillations. The ultimate period p_u is equal to the steady oscillation period. Then, the PID control is written as

$$K_\mathrm{PID} = K_\mathrm{P}\left(1 + \frac{1}{T_\mathrm{I}s} + T_\mathrm{D}s\right),\tag{5.2}$$

where K_P can be set to $0.6k_u$, T_I can be set to $p_u/2$, and T_D can be set to $p_u/8$ according to the modified Ziegler-Nichols tuning table [2].

In this section, we apply the PID tuning method to a piezo-electric smart system. Figure 5.1 shows the experimental tracking performance at 25 Hz. The maximum error of the output displacement is up to 2 μm. The error is 5% of the reference amplitude. As the input frequency increases, the tracking performance is degraded. Typically, PID tuning feedback control is suitable for moderate control or precision tracking at very low frequencies, such as 0.1 Hz.

Moreover, in this section we investigate the PID tuning by increasing the control gain. Figure 5.2 shows the PID tracking performance with the control gain increased by 30%. It can be seen that there are oscillations. Figure 5.3 shows the PID tracking performance with the control gain increased by 40%. It can seen that the oscillations become worse. This chattering indicates that the tracking performance of feedback control is limited by the high-frequency modes. To avoid oscillations and chattering, the feedback control should be limited to a safe range.

Finally, the PID tracking control performance at 300 Hz is presented in Figure 5.4. It can be seen that the output displacement of the piezoelectric system is less than 70% (−3 dB) of the reference signal. The feedback control has lost the bandwidth. This indicates that PID tuning control is not suitable for tracking a high-frequency signal. To expand the tracking bandwidth to high frequencies, feedforward control will be used as an alternative.

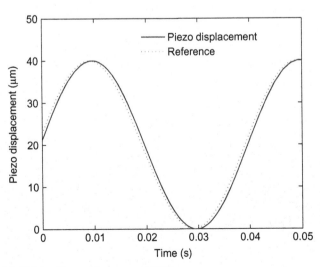

Figure 5.1 PID tracking performance at 25 Hz.

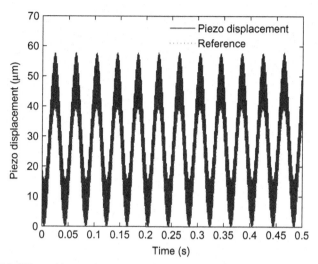

Figure 5.2 PID tracking performance with the control gain increased by 30%.

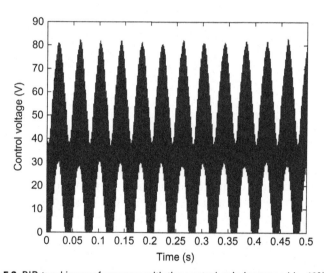

Figure 5.3 PID tracking performance with the control gain increased by 40%.

5.2.2 Ziegler-Nichols Tuning of Systems with Hysteresis

The PID tuning in smart systems is different from that in ordinary nonhysteretic systems, since the hysteresis has a special response under the input voltage. More details can be found in Ref. [3]. Figure 5.5 shows illustrates relay-based PID tuning. Γ denotes the static hysteresis, and G_e and G_v represent the electrical and mechanical dynamics, respectively.

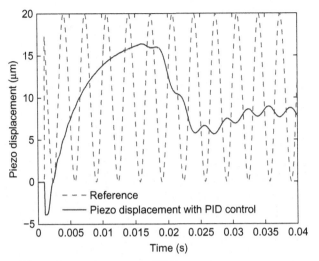

Figure 5.4 PID tracking performance at 300 Hz.

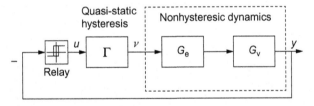

Figure 5.5 PID tuning of smart actuators.

The identified model of smart systems is used to compute the ultimate gain and ultimate period, and Ziegler-Nichols or other PID tuning rules can be used to obtain the parameters of the PI controller. First, the ultimate gain and period are computed with the identified nonhysteretic dynamics. Then, the ultimate gain is adjusted with respect to the Preisach model, and the ultimate period is kept unchanged as the Preisach model does not affect it.

Piezoelectric actuators have hysteresis and electrical and vibration dynamics. The hysteresis is represented by the Preisach model, which has a static and path-dependent nonlinearity. The static hysteresis does not exhibit a dynamic response, but will affect the input gain. Conversely, the electrical and vibration dynamics exhibit dynamic responses. Thus, standard PID tuning methods using a step response or relay tuning result in uncertainty of the ultimate gain.

The gain uncertainty due to Preisach hysteresis at input voltage $u(t)$ is illustrated in Figure 5.6. $S^{\max}$ is the area corresponding to

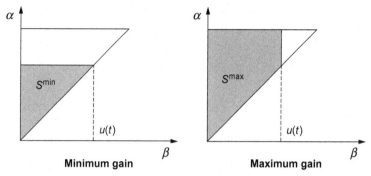

Figure 5.6 Area of uncertain gain in the Preisach plane.

the maximum gain at $u(t)$. Conversely, S^{min} is the area corresponding to the minimum gain at $u(t)$.

The minimum and maximum gains of input $u(t)$ due to
Preisach hysteresis are represented as

$$\Delta_{max} = \iint_{S^{max}} \mu(\alpha, \beta)\gamma(u(t))\, d\alpha\, d\beta, \tag{5.3}$$

$$\Delta_{min} = \iint_{S^{min}} \mu(\alpha, \beta)\gamma(u(t))\, d\alpha\, d\beta. \tag{5.4}$$

Thus, the gain uncertainty Δ due to Preisach hysteresis is represented as $\Delta_{min} < \Delta < \Delta_{max}$.

The ultimate period p_u and the gain margin g_m are determined
at the cross frequency ω_c of the nonhysteretic dynamics G:

$$p_u = \frac{1}{\omega_c}, \tag{5.5}$$

where ω_c is the cross frequency and is on the order of hertz.
The gain margin g_m of the nonhysteretic dynamics at the cross
frequency ω_c is given by

$$g_m = \left\| \frac{1}{G(j\omega)} \right\|, \tag{5.6}$$

where $G(j\omega)$ is the gain of the nonhysteretic dynamics at ω_c.

Then, the overall ultimate gain k_u is obtained by adjustment of
g_m with the Preisach gain:

$$k_u = \frac{g_m}{\Delta_{max}}. \tag{5.7}$$

Finally, the PI controller according to the Ziegler-Nichols tuning rule is represented by

$$K_{PI} = 0.6k_u \left(1 + \frac{2}{\tau_i} \frac{T}{2} \frac{z+1}{z-1} \right),$$

(5.8)

where T is the sampling interval, and k_u and τ_i are the ultimate gain and ultimate period, respectively.

5.2.3 Integral Control

In experiments with piezoelectric smart systems, the measured response time is on the order of milliseconds. Thus, in the solved PID controller, the integral gain is about 1000 times larger than the proportional gain and is more than 1 million times larger than the deferential gain. Thus, only integral feedback can be used to control piezoelectric smart systems. In such integral control, the measurement noise is reduced, but the closed-loop bandwidth is also reduced compared with that in the full PID tuning controller.

With only the proposed integral component in (5.8), Figure 5.7 shows the tracking performance for a sinusoidal signal at 10 Hz. The peak-peak amplitude of the tracking error is less than 5% of the peak-peak amplitude of the desired sinusoidal signal. Figure 5.8 shows the tracking performance for a triangular signal at 10 Hz. The peak-peak amplitude of the tracking error is 6.5% of the peak-peak amplitude of the desired triangular signal. The experimental results demonstrate the effectiveness of integral control.

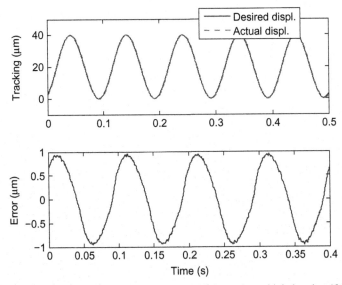

Figure 5.7 Tracking error (root mean square 0.66) for a sinusoidal signal at 10 Hz.

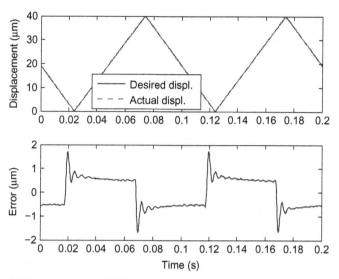

Figure 5.8 Tracking error at 10 Hz.

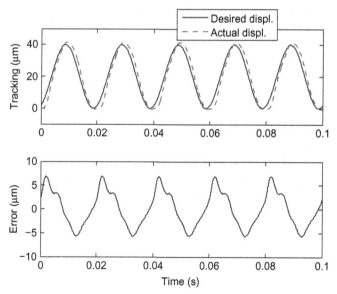

Figure 5.9 Tracking error (root mean square 3.8) for a sinusoidal signal at 50 Hz.

As the reference signal frequency increases to 50 Hz, the tracking performance is degraded. Figure 5.9 shows the tracking performance for the sinusoidal signal at 50 Hz. The peak-peak amplitude of the tracking error is 32.5% of the peak-peak amplitude of the desired sinusoidal signal. This indicates that integral control is more suitable for tracking a low-frequency signal.

5.3 Inversion-Based Feedforward Control

Inversion-based feedforward control is investigated in this section and was reviewed in Refs. [4, 5]. First, static Preisach hysteresis-based inversion-based feedforward control is presented. Then, composite-representation-hysteresis-based inversion-based feedforward is presented. Additionally, inversion-based feedforward control is applied to a piezoelectric system.

5.3.1 Preisach Hysteresis-Based Feedforward Control

To apply Preisach hysteresis-based feedforward control at low frequencies, let the hysteresis output f denote the output of smart systems. Then, Preisach hysteresis-based feedforward control is used. Figure 5.10 shows a flowchart for Preisach hysteresis-based inversion-based feedforward control [6]. u_{ff} denotes the feedforward voltage, f denotes the hysteresis output, and x_{r} denotes the desired value. Figure 5.11 shows the tracking performance for a sinusoidal signal at 25 Hz. H represents the static Preisach hysteresis. It can be seen that Preisach hysteresis-based feedforward control is effective in the experiment, but the tracking performance is limited owing to the modeling error, and there

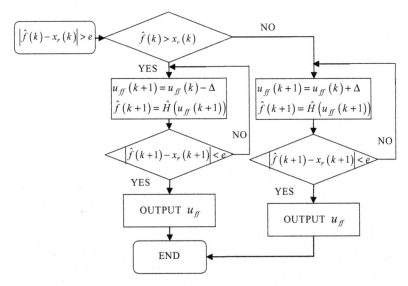

Figure 5.10 Flow chart for Preisach hysteresis-based inversion-based feedforward control.

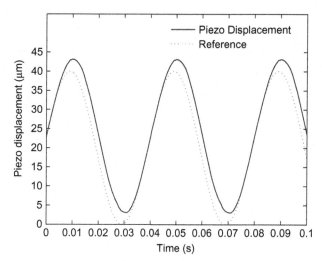

Figure 5.11 Preisach-based feedforward control of the sinusoidal signal at 25 Hz.

is an offset problem. To overcome these problems, composite hysteresis-based feedforward control is alternatively investigated.

5.3.2 Composite Hysteresis-Based Feedforward Control

Composite hysteresis-based feedforward control uses mechanical vibration, electrical dynamics, the static hysteresis effect, and the creep effect. The dynamics and effects will be compensated at broadband frequencies. Figure 5.12 shows inversion-based feedforward control using mechanical vibration, electrical dynamics, static hysteresis, and the creep effect. $\hat{G}_c$ denotes the estimation of the linear creep effect G_c, $\hat{G}_{ev}$ denotes the estimation of the mechanical vibration and electrical dynamics G_{ev}, Γ denotes the static hysteresis, x_r represents the reference or desired displacement, and y represents the output displacement.

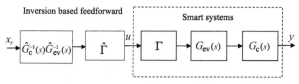

Figure 5.12 Inversion-based feedforward control using mechanical vibration, electrical dynamics, static hysteresis, and the creep effect.

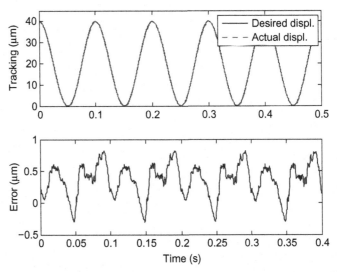

Figure 5.13 Inversion-based feedforward control of a sinusoidal signal at 10 Hz.

Figure 5.13 shows the tracking performance of the static hysteresis and creep based feedforward control. The peak-peak amplitude of the tracking error is less than 3%.

Figure 5.14 illustrates the tracking performance of a piezoelectric smart system obtained with use of the inversion of the mechanical vibration, electrical dynamics, static Preisach hysteresis, and the creep effect. The inversion-based feedforward control

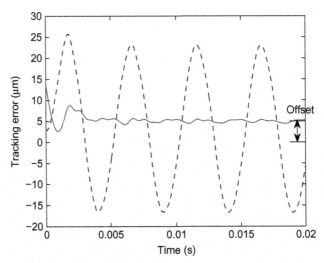

Figure 5.14 Inversion-based feedforward control of a sinusoidal signal at 200 Hz (the dashed line represents the open-loop error without compensation and the solid line represents the inversion-based feedforward control error).

error is smaller than the open-loop error without compensation, but there is still an offset value. The composite hysteresis-based inversion-based feedforward control is effective at high frequencies.

5.4 Composite Control

Composite control comprises inversion-based feedforward control and feedback control. Inversion-based feedforward control is used to improve the tracking bandwidth. The disturbances and offset are compensated by means of feedback control. First, we combine the inversion-based feedforward control in Section 5.3.2 and the PID control in Section 5.2. Figure 5.15 shows the composite control performance obtained by combination of the inversion-based feedforward control and PID tuning feedback control. Figure 5.16 shows the tracking errors of the Preisach-based inversion-based feedforward control, PID feedback control, and composite control. The proposed composite control gives the best performance among the three controllers. Moreover, Figure 5.17 shows the control voltage of the composite control. Finally, Figure 5.18 illustrates the input-output loops with the feedforward control, feedback control, and composite control. The proposed composite control suppressed the hysteresis loop effectively.

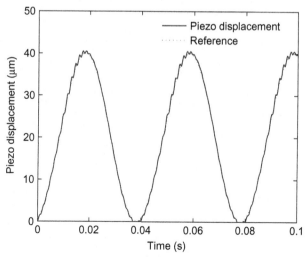

Figure 5.15 Composite control performance obtained through combination of the inversion-based feedforward control and PID feedback control.

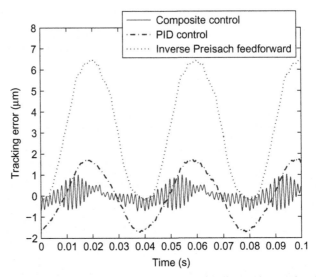

Figure 5.16 Tracking errors of the inversion-based feedforward control and PID feedback control.

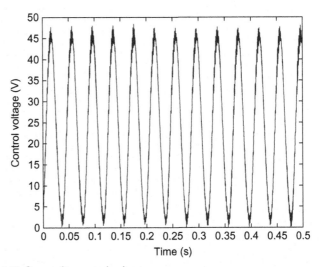

Figure 5.17 Composite control voltage.

5.5 Multirate Composite Robust Control

In this section, composite control is expanded to the case of multirate and robust control. High-bandwidth precision tracking of smart systems requires fast updating of control signals, especially at frequencies several times the first resonant frequency.

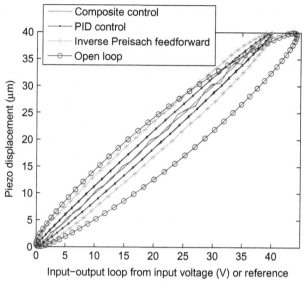

Figure 5.18 Input-output loops with the feedforward control, feedback control, and composite control.

However, most modern high-performance controllers are complex and computationally intensive, which restricts the update rates of control signals. In this section we present a multirate composite robust controller consisting of a slow-sampling H_∞ feedback controller and a fast-sampling feedforward controller for high-bandwidth tracking. The feedback controller is designed for stability and robustness in the presence of modeling errors and disturbances. The feedforward controller is designed for fast precision tracking. The proposed composite controller is realized in a real piezoelectric actuator based on a dSPACE 1104 board. With 1 kHz sampling in the feedback loop and 40 kHz sampling in the feedforward branch, the root mean square (RMS) tracking error at 1 kHz (twice times the ultimate frequency) is less than 2.3% of the reference amplitude. The feedforward is sampled as fast as possible. The feedback bandwidth is limited. This is a trade-off to reject disturbance and degrade feedforward performance

Currently, high-speed nanofabrication and high-speed imaging of scanning probe microscopes based on piezoelectric actuators are finding increasing applications, requiring piezoelectric actuators to perform beyond the first resonant frequency. However, classical controllers have limited performance at high frequencies. Moheimani [7], Clayton et al. [4], Devasia et al. [8], and Mahmood et al. [9] investigated the controllers of piezoelectric scanners and concluded that commercial atomic force microscopes typically operate at frequencies lower than 10% of the first resonant frequency because of complex hysteresis, creep, and electrical and vibration dynamics. To avoid mechanical vibrations, in most published work to the best of the authors' knowledge, the fine scanning of piezoelectric actuators currently achieved is typically limited to less than the resonant frequency

To compensate hysteresis dynamics and achieve higher-bandwidth tracking, various model-based modern controllers were investigated. Clayton et al. [4] reviewed feedforward approaches that are mainly based on a linear dynamics model. Wu and Zou [10] presented a feedforward-feedback controller with two degrees of freedom and reported a tracking error of 5.28% at 300 Hz (lower than the first resonant frequency). Leang and Devasia [11] proposed a notch filter and inversion-based feedforward to enhance the high-gain feedback. However, the hysteresis was not modeled and fine tracking at frequencies higher than the resonant frequency was still not achieved in the smart systems experiment. Intelligent feedback controllers were also investigated. Liaw and Shirinzadeh [12] presented a neural network to enhance the motion tracking of piezoelectric-based flexible mechanisms. Adaptive controllers are proposed in Ref. [13]. Additionally, dynamic hysteresis models were investigated to achieve high-bandwidth tracking [14]. On the basis of rate-dependent Prandtl-Ishlinskii hysteresis, Tan et al. [15] proposed hysteresis-based inversion to extend the tracking bandwidth, but it is difficult to apply modern control techniques to rate-dependent hysteresis, since most modern controllers are designed with nonhysteretic models.

Advanced and intelligent feedback controllers are commonly complex and need more calculations, which limits the updating rates for feedback control signals. For instance, H_∞ and H_2 controllers may have very high orders, since they have the same order as the generalized plant [16]. In a real-time implementation, fast digital signal processors (DSPs) may be required. At high frequencies, these complex controllers are not adequate because of hardware limitations. They are typically used to track the reference trajectories lower than the resonant frequency [17], because of potential dangers of large gain at high frequencies.

Multirate composite controllers can also achieve high-speed tracking. Although various multirate composite controllers have been investigated, the objectives and approaches are different from those in this chapter. For example, Fujimoto et al. [18, 19] presented perfect tracking control based on multirate feedforward to handle non-minimum phase zeros. Cao et al. [20] modeled the frequency-domain transfer function of a multirate controller to evaluate the open-loop response. Duan et al. [21] provided an adaptive feedforward based on multirate discretization to compensate periodic disturbances. Lee [22] proposed a multirate feedback controller to improve the intersample behavior. In addition, the hysteresis of smart systems is also not modeled.

In this section, the multirate controller is designed for limited DSPs to achieve fast and accurate tracking. The objective of this section is to design a multirate composite controller that can be used on slow DSPs. The controller can achieve accurate tracking at frequencies twice as high as the first resonant frequency of piezoelectric actuators. The feedback controller is designed to guarantee robust stability and disturbance suppression according to the disturbance characteristics and its sampling rate. The feedforward controller is designed to achieve fast tracking, based on the model inversion of the identified hysteretic dynamics of piezoelectric actuators.

5.5.1 Proposed Multirate Composite Control Strategy

In this section, the multirate composite controller is presented. It consists of a model-based feedforward controller and a discrete H_∞ feedback controller. First, the model-based inversion-based feedforward controller is constructed to compensate the phase lag and magnitude distortion. The reference is assumed to be known. To improve the computing efficiency, the feedforward control signals are computed off-line. Then, they are read by a DSP platform for real-time implementation. The discrete H_∞ feedback controller with integral action is built to suppress disturbances at low frequencies. The design of the H_∞ controller is a trade-off among the feedforward error, disturbances, calculations, and noise. The bandwidth and sampling rates are determined for the feedforward and feedback loops. The feedforward control is sampled with a fast sampling rate, since it can be computed off-line. The feedback control is sampled with a slow sampling rate compared with the off-line feedforward control.

Figure 5.19 illustrates the proposed composite control strategy. "ZOH" means zero-order-hold sampling, u_{ff} and u_{fb} are the feedforward and feedback signals, respectively, e, d, and e denote the tracking error, output disturbance, and measurement noise, G denotes the nonhysteretic dynamics in a piezoelectric actuator and $G = G_c G_{ev}$, f_{FF}, f_{FB}, and f_M denote the sampling rates of the feedforward, feedback, and measurement signals, respectively($f_{FF} = f_M$ is used in this chapter), $\hat{G}^{-1}$ and $\hat{\Gamma}^{-1}$ denote the inversion of

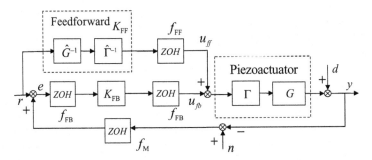

Figure 5.19 Multirate composite control strategy for piezoelectric actuators.

the estimated nonhysteretic dynamics and hysteretic dynamics, respectively, and K_{FF} and K_{FB} denote the feedforward controller and the feedback controller. The feedforward controller is constructed on the basis of the hysteresis and nonhysteretic dynamics. The discrete feedback controller is designed with use of the nonhysteretic dynamics, and the rate-independent hysteresis is regarded as a bounded uncertainty.

5.5.2 Analysis of the Proposed Composite Controller

The proposed composite controller without consideration of multirate is analyzed in this section. The model-based inversion-based feedforward controller K_{FF} of piezoelectric actuators can be represented as

$$K_{FF} = \hat{G}^{-1} \circ \hat{\Gamma}^{-1}, \tag{5.9}$$

where $\circ$ denotes the composition operator [23, 24]. Multiplication is not used in (5.9) since the hysteresis Γ and its estimation $\hat{\Gamma}$ are strong nonlinearities with global memories [25, 26]. $\hat{G}^{-1}$ and $\hat{\Gamma}^{-1}$ can be represented as

$$\begin{cases} \hat{G}^{-1} = G^{-1}(1 + \delta_l), \\ \hat{\Gamma}^{-1} = \Gamma^{-1}(1 + \delta_h), \end{cases}$$

where δ_l denotes the inversion error of the nonhysteretic dynamics G and δ_h denotes the inversion error of the rate-independent hysteresis Γ. δ_l and δ_h are bounded uncertainties and are determined by the identification accuracy of piezoelectric actuators.

Then, $\hat{\Gamma}^{-1}$ and $\hat{G}^{-1}$ can be rewritten as

$$\hat{\Gamma}^{-1} \circ \hat{G}^{-1} = \Gamma^{-1} \circ G^{-1}(1 + \delta_l + \delta_h + \delta_l \delta_h).$$

Let $\delta = \delta_l + \delta_h + \delta_l \delta_h$, and then the model-based inversion feedforward controller of piezoelectric actuators is represented as

$$K_{FF} = \Gamma^{-1} \circ G^{-1}(1 + \delta), \tag{5.10}$$

where the bounded uncertainty δ can be determined with use of feedforward control.

With only the feedforward controller K_{FF} in (5.10)—the feedback controller $K_{FB} = 0$—the relative error in e/r is given by

$$\frac{e}{r}\Big|_{K_{FB}=0} = \delta + \frac{d}{r}. \tag{5.11}$$

Equation (5.11) indicates that the tracking performance of feedforward control relies on the identification accuracy and the output disturbances cannot be suppressed. Thus, feedback control is necessary to guarantee the stability and robustness under modeling error δ and disturbance d.

Figure 5.20 shows the proposed composite control, where $\hat{G}^{-1}$ and $\hat{\Gamma}^{-1}$ are represented by G^{-1} and Γ^{-1}, respectively, according to (5.10). With the proposed composite control, the relationship between the reference r and the piezoelectric actuator displacement output y is written as

$$\frac{y}{r} = 1 + \frac{1}{G \circ \Gamma \circ K_{FB} + 1}\delta + \frac{G \circ \Gamma \circ K_{FB}}{G \circ \Gamma \circ K_{FB} + 1}\frac{n}{r} + \frac{1}{G \circ \Gamma \circ K_{FB} + 1}\frac{d}{r},$$

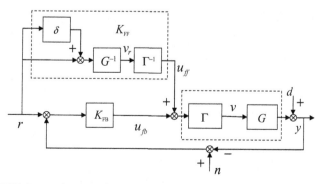

Figure 5.20 Composite control strategy.

where δ is the feedforward error, K_{FB} denotes the feedback controller, and n and d are the measurement noise and output disturbances, respectively.

Then, the relationship between the reference r and the tracking error e is represented as

$$\frac{e}{r} = \frac{1}{G \circ \Gamma \circ K_{FB} + 1}\delta + \frac{1}{G \circ \Gamma \circ K_{FB} + 1}\frac{d}{r} + \frac{G \circ \Gamma \circ K_{FB}}{G \circ \Gamma \circ K_{FB} + 1}\frac{n}{r}. \tag{5.12}$$

The feedback controller K_{FB} is designed to suppress the output disturbance d and the feedforward error δ, but the measurement noise n is amplified in the feedback bandwidth. Some signals in noise n may coincide with mode frequencies of piezoelectric actuators, resulting in chattering and unstable responses. Thus, multiobjective robust H_∞ control is used to design K_{FB}—that is, different objectives are specified at different frequencies for the composite controller. The sampling rate and bandwidth of K_{FB} are the main specifications to be considered.

Remark. The model-based inversion-based feedforward controller is used to expand the bandwidth beyond resonant frequencies of piezoelectric actuators, because noise lies outside the feedforward branch. Conversely, the feedback controller is used to reject disturbances and modeling error within the feedback bandwidth. The high frequency gain of the feedback controller should be reduced to suppress the noise effect.

5.5.3 Off-Line Model-Based Inversion-Based Feedforward Controller

An off-line model-based inversion-based feedforward controller can be used if the reference is known. The feedforward controller is used to overcome the bandwidth limitation of the feedback controller. It encompasses the inverse nonhysteretic dynamics and the inverse hysteresis. First, the reference signals pass through the inverse nonhysteretic dynamics $\hat{G}^{-1}$, and then the inverse hysteresis $\hat{\Gamma}^{-1}$. Details of the nonhysteretic dynamics inversion can be found in Refs. [27, 28].

With periodic trajectories, the inversion of the nonhysteretic dynamics of piezoelectric actuators can be represented by the steady response of $\hat{G}^{-1}$, which is represented by

$$\hat{G}^{-1}(s) = \hat{k}_{ev}(\hat{\tau}s + 1) \prod_{i=1}^{m} \frac{s + \hat{p}_i}{s + \hat{z}_i} \cdot \frac{\prod_i^n (s^2 + 2\hat{\xi}_i\hat{\omega}_i s + \hat{\omega}_i^2)}{\prod_j^{n-2}(s^2 + 2\hat{\xi}_j\hat{\omega}_j s + \hat{\omega}_j^2)}, \tag{5.13}$$

where $\hat{k}_{ev}$, $\hat{\tau}$, $\hat{\xi}_i$, $\hat{\xi}_j$, $\hat{\omega}_j$, and $\hat{\omega}_j$ are the identified parameters of the electrical and vibration dynamics, and $\hat{z}_i$ and $\hat{p}_i$ are the estimated zeros and poles of the creep dynamics. Negative z_i and ξ_j are not considered in (5.13).

The feedforward control signal u_{ff} can be computed off-line through (5.13) for a known and periodic reference. Then, the data are read by a DSP platform in real-time feedforward control.

5.5.4 Design of Bandwidth and Sampling Rate

The design of the bandwidth and sampling rate of the multi-rate composite controller comprises their design in feedback and feedforward loops, respectively. The design of the bandwidth and sampling rate in the feedback loop is the main subject here, since the bandwidth and sampling rate in the feedforward loop are easy to determine. The feedforward bandwidth is determined by the identification accuracy of piezoelectric actuator dynamics and the sampling rate can be set to the fastest one that a DSP allows.

Design of Bandwidth

The bandwidth of the feedback control is important to guarantee stability and suppress disturbances. The feedback bandwidth is designed according to the performance requirements, disturbance characteristics, and modeling uncertainty. Design of the bandwidth and sampling rate in the feedback loop is a trade-off among tracking errors, disturbances, and measurement noise. Let the bandwidth of output disturbance d be ω_d. The measurement noise n is assumed to be white noise. First, the maximum sampling rate f_{max} of a DSP platform is a limitation of both the feedforward and the feedback bandwidth [29]:

$$\omega_{FF} \leq f_{max}, \quad \omega_{FB} < \frac{f_{max}}{2}.$$

To reduce modeling error, offset, drift, and other disturbances in piezoelectric mechanisms, the bandwidth of feedback control is higher than the disturbance bandwidth ω_d:

$$\omega_{FB} > \gamma \omega_d,$$

where γ denotes the suppression performance of disturbances. γ should be as large as possible to reject more disturbances. The modeling error, drift, offset, and other disturbances lower than ω_d can be suppressed by K_{FB}.

The bandwidth of feedback control is also limited to the frequency ω_{bc} to guarantee robust stability under the modeling error

δ, the measurement noise n, and the coupling with feedforward control at high frequencies. ω_{bc} can be estimated by the Nyquist stability criterion:

$$\omega_{FB} < \omega_{bc}.$$

Additionally, the computations of feedback controller K_{FB} are another limitation. Let the maximum calculation time of K_{FB} be T_{FB}. T_{FB} increases as the order and complexity of K_{FB} increase. To guarantee the accurate updating of feedback control signals in a real-time implementation, the feedback bandwidth should satisfy

$$\omega_{FB} < \frac{1}{T_{FB}}.$$

Finally, the feedback bandwidth satisfies

$$\gamma \omega_d < \omega_{FB} < \min \left(\frac{f_{max}}{2}, \frac{1}{T_{FB}}, \omega_{bc} \right). \tag{5.14}$$

Design of the Sampling Rate in the Feedback Loop

The sampling rate in the feedback loop is related to the feedback bandwidth, the calculating capacity of DSPs, the noise level, and the ultimate frequency of a piezoelectric actuator. In the real-time control of the piezoelectric actuator experiment, too fast sampling in the feedback loop may degrade the performance, and even the stability might be lost under too fast sampling. First, a DSP needs time to calculate and update the feedback signal, but the limited calculating performance of a DSP restricts the sampling rate, especially for a complex and modern controller. Second, fast sampling of high-gain feedback controllers amplifies the high-frequency noise n, which may contains signals coinciding with resonant (mode) frequencies of piezoelectric actuators, resulting in serious chattering.

Both the upper bound and the lower bound of the sampling rate f_{FB} are necessary for the feedback controller K_{FB}. Let $\bar{f}_{FB}$ denote the upper bound of sampling rates that can be implemented in DSPs. $\bar{f}_{FB}$ is mainly related to the calculations of the feedback controller K_{FB}. Additionally, the lower bound of sampling rate $\underline{f}_{FB}$ is related to the disturbance bandwidth ω_d, the ultimate frequency ω_r of the piezoelectric actuator, and the feedback bandwidth ω_{FB}. Thus, the sampling rate f_{FB} in the feedback loop is given by

$$\underline{f}_{FB} < f_{FB} < \bar{f}_{FB}, \tag{5.15}$$

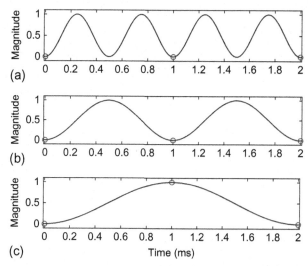

Figure 5.21 Slow sampling rate of signals (the solid line represents reference signals and the circles represent sampling points).

where $f_{\underline{FB}} = 2 \max(\omega_{FB}, \omega_r, \omega_d)$ and $\bar{f}_{FB} = \min\left(f_{max}, \frac{1}{T_{FB}}\right)$.

Figure 5.21 illustrates a slow sampling rate in the feedback loop. The frequencies of the reference trajectories in Figure 5.21 are 200%, 100%, and 50% of the sampling rate.

Remark. Compared with the sampling in the feedforward loop, insufficient sampling can be used for the feedback controller when the trajectory rates are beyond the resonant frequencies of piezoelectric actuators. In this case, the feedback controller still rejects disturbances within its bandwidth, but has no tracking performance.

5.5.5 Design of a Discrete Feedback Controller

This section presents the design of a discrete H_∞ controller according to the sampling rate f_{FB} and the disturbance bandwidth ω_d. Additionally, a loop-shaping technique is used. The feedforward controller is effective in reducing phase lag and achieving high-bandwidth tracking, but there is an undesirable modeling error and other disturbances. It is necessary to design an optimal feedback controller to reject disturbances and guarantee the robustness of stability and performance.

Figure 5.22 illustrates the loop-shaping technique used to design the feedback controller. The performance and stability requirements are satisfied by specification of L_1 and L_2. ω_c is the cross frequency of GK_{FB} and is related to and close to the feedback

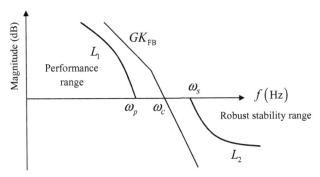

Figure 5.22 Loop shaping considering disturbance and sampling rate.

bandwidth ω_{FB}, ω_p is related to the disturbance rejection performance, and ω_s is related to the robust stability under disturbances and measurement noise at high frequencies. ω_p and ω_s can be represented as

$$\omega_p = \gamma\omega_d, \quad \omega_s = \min(\frac{\omega_{max}}{2}, \frac{1}{T_{FB}}, \omega_{bc}). \tag{5.16}$$

Weighting functions are suitable to specify different requirements at different frequencies as in Figure 5.22. It is convenient to achieve multiple objectives with weighting functions [16]. The robust discrete H_∞ controller is designed on the basis of nonhysteretic dynamics, and the rate-independent hysteresis Γ can be regarded as an input uncertainty consisting of the nominal gain k_h and the weighting function w_u.

Figure 5.23 shows a sketch of multiobjective robust H_∞ control. w_1 is the performance weighting function to specify the performance requirements and achieve fine tracking. Moreover, significant vibrations are easily induced by high gain at high frequencies. Then, integral action is added to w_1 to reduce the feedback bandwidth ω_{FB} and enhance the disturbance suppression at low frequencies. w_n, w_r, and w_d denote the noise, reference, and disturbance weighting functions, respectively, w_2 is the control weighting function to limit the control gain and suppress noise at high frequencies, w_u denotes the uncertainty due to the hysteresis nonlinearity, and Δ_u is the unit complex uncertainty with norm $\|\Delta_u\| < 1$.

The inverse error δ of the feedforward control can be written as

$$\delta = w_\delta \Delta_\delta, \tag{5.17}$$

where Δ_δ denotes a unit complex uncertainty with $\|\Delta_\delta\| \leq 1$, and w_g is the weighting function to describe the feedforward error.

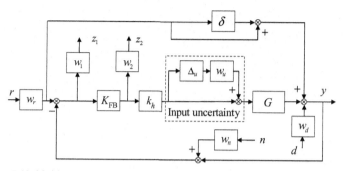

Figure 5.23 Multiobjective H_∞ control.

Weighting functions w_1 and w_2 are used to satisfy the trajectory of GK_{FB} that is bounded by L_1 and L_2. The relationship is as follows:

$$\begin{cases} w_1|_{\omega \leq \omega_p} = L_1, \\ w_2|_{\omega \geq \omega_s} = 1/L_2. \end{cases} \tag{5.18}$$

G is the nonhysteretic dynamics of the piezoelectric actuator, and its identification $\hat{G}$ can be regarded as the nominal model to design the discrete H_∞ controller. The discrete state space of the nominal model can be represented as

$$\begin{cases} x(k+1) = \bar{A}x(k) + \bar{B}v(k), \\ y(k) = \bar{C}x(k), \end{cases} \tag{5.19}$$

where $v(k)$ is the output of rate-independent hysteresis, abd $\bar{A}$, $\bar{B}$, and $\bar{C}$ can be transformed from the identified nonhysteretic dynamics in piezoelectric actuators. The sampling rate f_{FF} in the feedforward loop is N times f_{FB} in the feedback loop (i.e., $f_{FF} = Nf_{FB}$).

Remark. The specifications and the feedforward control error are incorporated into the feedback controller by loop-shaping techniques.

5.5.6 Experimental Setup

The experimental setup consists of a piezoelectric stage, an amplifier, a linear variable differential transformer (LVDT), and a dSPACE 1104 board. Figure 5.24 shows the piezoelectric stage. The stage has a travel span of 80 μm. The low-voltage lead zirconate titanate amplifier is an E-662 amplifier with a output voltage range of $[-20, 120]$ V. The hysteretic dynamics of the piezoelectric stage was identified in Refs. [3, 28]. The piezoelectric

Figure 5.24 Piezoelectric stage in the experiment.

stage is shown in Figure 5.24. It is a smart system with high stiffness. The identification approaches in Chapter 4 are used, and the identified results are summarized in this section.

The creep dynamics is identified as

$$\hat{G}_c(s) = \frac{(s + 0.01458)(s + 0.1716)(s + 0.241)}{(s + 0.01419)(s + 0.1684)(s + 0.2402)}$$
$$\cdot \frac{(s + 1.07)(s + 18.29)}{(s + 1.053)(s + 17.57)}. \tag{5.20}$$

The electrical and vibration dynamics are modified and identified as

$$\hat{G}_{ev}(s) = \frac{1}{0.000474s + 1} \cdot \frac{8.111 \times 10^6}{s^2 + 3786s + 8.111 \times 10^6}$$
$$\cdot \frac{2.478 \times 10^7}{s^2 + 809.1s + 2.478 \times 10^7}. \tag{5.21}$$

The identified density function μ is shown in Figure 5.25.

5.5.7 Proposed Multirate Composite Controller

This section presents the multirate composite controller of the piezoelectric stage obtained with the design strategy in Sections 5.5.3-5.5.5. The model-based inversion-based feedforward is achieved first. The nonminimum zeros of the identified nonhysteretic dynamics are 3 times faster than the stable zeros. Thus, the nonminimum zeros are ignored to guarantee the stability of the model-based inversion-based feedforward. In addition, the

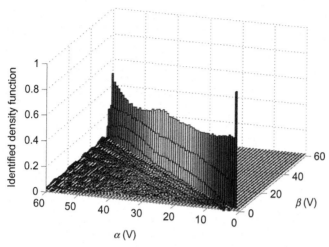

Figure 5.25 Identified density function μ.

maximum sampling rate f_{max} of the dSPACE 1104 board is 40 kHz. Then, the sampling rate f_{FF} in the feedforward loop is 40 kHz.

After the feedforward performance has been tested, the percent feedforward-error function δ, without considering the offset and drifts in the displacement y, is less than 5%. Thus, $w_\delta = 0.05$. On the basis of the identified Preisach hysteresis, the nominal gain of the hysteresis is $k_h = 0.96$, and the weighting function reflecting the input uncertainty is $w_u = 0.12$. The disturbances such as the displacement offset and drifts are at frequencies below 10 Hz, indicating that the disturbance bandwidth $\omega_d = 10$ Hz. To reject more disturbances, $\gamma = 2$ and $w_p = 20$ Hz are used in the experiment. The ultimate frequency of the piezoelectric stage is 500 Hz (i.e., $\omega_r = 500$ Hz). To reduce high-frequency noise and induced vibrations, $w_s = 350$ Hz.

The sampling rate f_{FB} is flexible to design. In this chapter, a much slower f_{FB} is tested, such that the proposed multirate controller can be used in a slow DSP. The lower bound due to the feedback bandwidth, ultimate frequency, and disturbance bandwidth in (5.15) is 1 kHz.

According to w_p and w_s, with a sampling rate $f_{FB}=1$ kHz in the feedback loop, the weighting functions w_1 and w_2 used in the H_∞ control are as follows:

$$\begin{cases} w_1 = 0.05(z+1)/(z-1), \\ w_2 = 1.2496(z-0.2283)^2/(z+0.7254)^2, \end{cases}$$

where $(z+1)/(z-1)$ is the integral action of w_1.

With the same f_{FB}, the reference signal, disturbances, and measurement noise are represented with the weighting functions w_r, w_d, and w_n, respectively:

$$\begin{cases} w_r = 1, \\ w_d = 0.09138(z+1)/(z-0.9391), \\ w_n = 0.0395(z-0.9691)/(z+0.222). \end{cases}$$

Doyle et al. [30] presented the analytical solution of H_∞ optimization, but the optimization through the H_∞ norm is conservative. To reduce the conservation, discrete D-K iteration with a structured singular value is used to solve the controller [16]. With use of the weighting functions, the structure singular value is 0.907 after four iterations. The robust stability is satisfied according to the small-gain theorem. The H_∞ controller is as follows:

$$K_{FB}(z) = \prod_{j=1}^{k} \frac{z + z_{kj}}{z + p_{kj}}, \tag{5.22}$$

where k is the order of the controller, $K = 15$. The zeros of $K_{FB}(z)$ are $-1, -0.6836 \pm 0.6529i, -0.7254, -0.2088 \pm 0.4324i, -0.2220, -0.0266, 0.6938, 0.8603, 0.9170, 0.9825, 0.9989, 0.9998,$ and 1.0000, and the poles of $K_{FB}(z)$ are $-0.6809 \pm 0.6371i, -0.1156 \pm 0.5543i, -0.2598 \pm 0.3603i, -0.2116 \pm 0.0469i, 0.6862, 0.9134, 0.9946 + 0.0091i, 0.9970, 0.9997, 0.9998,$ and 1.0000.

5.5.8 Performance of the Proposed Multirate Composite Controller

To implement the feedback H_∞ controller easily, the reduced sixth-order H_∞ controller of the H_∞ controller in (5.22) is used.

For easy implementation, the harmonic signal $r(t)$ is chosen as the reference trajectory and is given by

$$r(t) = A \frac{1 - \cos(2\pi ft)}{2},$$

where A is the amplitude of reference trajectory $r(t)$, t is time, and f is the reference frequency.

The percent RMS error e_{RMS} is used to measure the tracking performance and is defined as [11]

$$e_{RMS}(\%) = \left(\frac{\sqrt{\frac{1}{l} \sum_{i=1}^{l} (r(i) - y(i))^2}}{\max(r) - \min(r)} \right) \times 100\%, \tag{5.23}$$

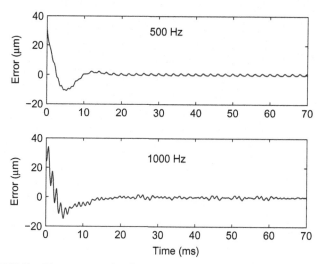

Figure 5.26 Tracking errors under disturbances.

where l is the number of sampling points, and $r(i)$ and $y(i)$ are the reference and piezoelectric displacement at time instant i.

To verify the effectiveness of the proposed multirate composite controller, the tracking performances at 500 Hz (ultimate frequency of the piezoelectric actuator) and 1000 Hz are investigated. The magnitudes of $r(t)$ at frequencies of 500 and 1000 Hz are 21 and 10.25 µm, respectively. The sampling rates in the feedback and feedforward loops are 1 and 40 kHz, respectively.

Transient Responses under Disturbances

Figure 5.26 shows the transilient responses of tracking errors obtained with the proposed multirate composite controller with the specified $r(t)$ at 500 and 1000 Hz, respectively. The tracking errors are reduced to the steady values in 20 ms. Thus, the performance of disturbance suppression of the slow-sampling feedback controller is verified. The corresponding steady responses are presented in the next section.

Steady Responses

The steady responses of the piezoelectric actuator at 500 and 1000 Hz are presented. Figure 5.27(a) and (b) shows the tracking performance at 500 Hz: the RMS tracking error is 0.486 µm and its e_{rms} in (5.23) is 2.3%. Figure 5.27(c) and (d) shows the feedforward and feedback control signals. In each period, 80 sampling points of the feedforward signals, which are computed off-line, are used to track the reference, but only two points of the reference

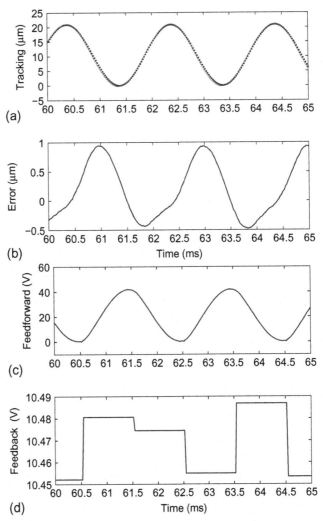

Figure 5.27 Tracking performance and control signals at 500 Hz.

and the measured displacement are used for feedback to reject disturbances.

Figure 5.28(a) and (b) shows the tracking performance at 1000 Hz (twice the ultimate frequency): the RMS tracking error is 0.226 μm and the e_{rms} in (5.23) is 2.2%. Figure 5.28(c) and (d) shows the feedforward and feedback control signals in volts. In each period, 40 sampling points of the feedforward signals are used to track the reference, but only one point of the reference and the measured displacement is used for feedback to reject disturbances.

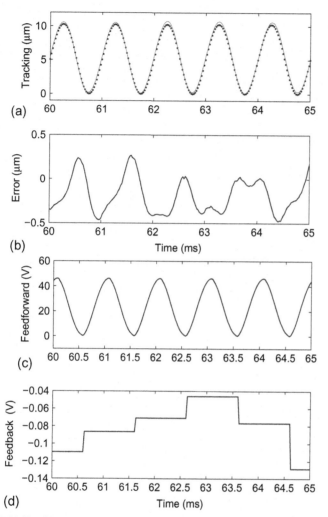

Figure 5.28 Tracking performance and control signals at 1000 Hz.

Figure 5.29 shows the power spectral density (PSD) of the feedforward and feedback control signals at 500 and 1000 Hz, respectively. The PSD of the feedback control signal is less than 100 Hz, since the H_∞ feedback controller is designed to suppress modeling error and disturbances at low frequencies and its bandwidth is less than the resonant frequency. The PSD of the feedforward control signals is largest at the reference frequencies—that is, 500 and 1000 Hz, respectively.

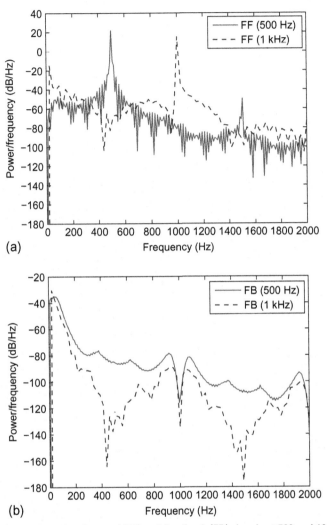

Figure 5.29 PSD of feedforward (FF) and feedback (FB) signals at 500 and 1000 Hz.

The control signals and their PSDs in Figures 5.27(c) and (d), 5.28(c) and (d), and 5.29 indicate that the feedforward control achieves fast tracking at high frequencies and the feedback control rejects disturbances at low frequencies.

Remark. The experimental studies of a piezoelectric stage have demonstrated the effectiveness of the proposed composite controller. The feedforward controller mainly contributes to the high-bandwidth tracking. In contrast, the robust feedback controller sufficiently rejects disturbances and modeling errors within the feedback bandwidth.

5.6 Control Approach Suggestions for Smart Systems with Hysteresis

In this section, the control approaches for smart systems with hysteresis are presented. We try to provide some suggestions for controller selection as follows.

PID-based classical control. PID-based classical control is effective for most smart systems. Precision tracking can be provided at low frequencies. Relay-based PID tuning is easily implemented. It is suggested that engineers choose PID tuning control for modest specifications or low-frequency precision motion.

Hysteresis-based feedforward control. Hysteresis-based feedforward control is affected by the measurement noise. In some situations where the positioning accuracy is sensitive to measurement noise, hysteresis-based feedforward control can be used, but there is an offset and disturbances. Additionally, feedforward control relies on accurate modeling of smart systems.

Nonhysteretic dynamics-based modern control. Nonhysteretic dynamics-based modern control is widely investigated. Engineers skilled in modern control theory can choose various modern control approaches. Generally, mechanical vibrations and electrical dynamics can represent smart systems with input uncertainty. For piezoelectric smart systems, the typical input uncertainty is 5-15%.

Nonhysteretic dynamics-based intelligent control. Most smart systems have no friction or other time-invariant nonlinearity. The responses of smart systems are very repeatable. Thus, intelligent control methods are suitable for a periodic reference, especially repetitive control. Engineers skilled in intelligent control are encouraged to use this type method in situations where there is a periodic reference.

Composite control. In composite control, the various feedforward control approaches and feedback control approaches can be synthesized. Composite control can exhibit precision and high bandwidth tracking simultaneously. Most composite control approaches are also complex. Composite control is a synthesis method.

5.7 Conclusion

To achieve fast and accurate tracking with limited DSPs, the multirate composite controller was presented. Fast-sampling feedforward control was built with use of the identified hysteretic dynamics. The slow-sampling discrete H_∞ controller was designed with use of the nonhysteretic dynamics. The proposed composite controller was implemented in a piezoelectric stage with a DSP platform. The percent RMS tracking errors at 500 and 1000 Hz were less than 2.2% and 2.3%, respectively. Fast and accurate tracking performance is verified.

References

[1] K.J. Astrom, T. Hagglund, Pid Controllers: Theory, Design and Tuning 2nd ed., Instrument Society of America, 1995.

[2] D.T. Wilson, Relay-based PID tuning, Automat. Control (2005) 3: 10-12.

[3] L. Liu, K.K. Tan, C.S. Teo, S.L. Chen, T.H. Lee, Development of an approach towards comprehensive identification of hysteretic dynamics in piezoelectric actuators, IEEE Trans. Control Syst. Tech. 2012, doi:10.1109/TCST.2012.2200896.

[4] G.M. Clayton, S. Tien, K.K. Leang, Q. Zou, S. Devasia, A review of feedforward control approaches in nanopositioning for high-speed SPM, ASME J. Dyn. Syst. Measur. Control 131 (2009) 0611011-06110119.

[5] K. Leang, Q. Zou, S. Devasia, Feedforward control of piezoactuators in atomic force microscope systems, IEEE Control. Syst. Mag. 29 (1) (2009) 70-82.

[6] L. Liu, K.K. Tan, S.L. Chen, S. Huang, T.H. Lee, SVD-based Preisach hysteresis identification and composite control of piezo actuators, ISA Trans. 51 (3) (2012) 430-438.

[7] S.O. Moheimani, Accurate and fast nanopositioning with piezoelectric tube scanners: emerging trends and future chanllenges, Rev. Sci. Instrum. 79 (2008) 0711011.

[8] S. Devasia, E. Eleftheriou, S. Moheimani, A survey of control issues in nanopositioning IEEE Trans. Control Syst. Tech. 15 (5) (2007) 802-823.

[9] I.A. Mahmood, S.R. Moheimani, B. Bhikkaji, A new scanning method for fast atomic force microscopy, IEEE Trans. Nanotechnol. 10 (2) (2011) 203-216.

[10] Y. Wu, Q. Zou, Robust inversion-based 2-Dof control design for output tracking: piezoelectric-actuator example, IEEE Trans. Control Syst. Technol. 17 (5) (2009) 1069-1082.

[11] K.K. Leang, S. Devasia, Feedback linearized inverse feedforward for creep, hysteresis and vibration compensation in AFM piezoactuators, IEEE Trans. Control Syst. Technol. 15 (5) (2007) 927-35.

[12] H. Liaw, B. Shirinzadeh, Neural Network motion tracking control of piezo actuated flexure based mechanisms for micro nanomanipulation, IEEE/ASME Trans. Mechatron. 14 (5) (2009) 517-527.

[13] H.J. Shieh, C.H. Hsu, An adaptive approximator-based backstepping control approach for piezoactuator-driven stages, IEEE Trans. Ind. Electron. 55 (4) (2008) 1729-1738.

[14] M. Janaiden, C. Su, S. Rakheja, Development of the rate-dependent Prandtl-Ishlinskii model for smart actuators, Smart Mater. Struct. 17 (2008) 035026.

[15] U.X. Tan, W.T. Latt, C.Y. Shee, C.N. Riviere, W.T. Ang, Feedforward controller of ill-conditioned hysteresis using singularity-free Prandtl-Ishlinskii model, IEEE/ASME Trans. Mechatron. 14 (5) (2009) 598-605.

[16] S. Skogestad, I. Postlethwaite, Multivariable Feedback Control: Design and Analysis, Wiley, New York, 2005.

[17] A. Fleming, Nanapositioning system with force feedback for high performance tracking and vibration control, IEEE/ASME Trans. Mechatron. 15 (3) (2010) 433-446.

[18] H. Fujimoto, B. Yao, Multirate adaptive robust control for discrete-time non-minimum phase systems and application to linear motors, IEEE/ASME Trans. Mechatron. 10 (4) (2005) 371-377.

[19] K. Saiki, A. Hara, K. Sakata, H. Fujimoto, A study on high-speed and high-precision tracking control of large-scale stage using perfect tracking control method based on multirate feedforward control, IEEE Trans. Ind. Electron. 57 (4) (2010) 1393-1400.

[20] W. Cao, Q. Bi, X. Liu, C. Lim, Y. Soh, Frequency domain transfer function of digital multirate controller with current estimator, IEEE Trans. Control Syst. Technol. 13 (1) (2005) 131-137.

[21] C. Duan, G. Gu, C. Du, T.C. Chong, Robust compensation of periodic disturbances by multirate control, IEEE Trans. Magn. 44 (3) (2008) 413-418.

[22] S.H. Lee, Multirate digital control system design and its application to computer disk drives, IEEE Trans. Control Syst. Technol. 14 (1) (2006) 124-133.

[23] M. Brokate, J. Sprekels, Hysteresis and Phase Transitions, Springer-Verlag, Berlin, 1996.

[24] C. Visone, Hysteresis modeling and compensation for smart sensors and actuators, J. Phys. Conf. Ser. 012028 (2008) 138.

[25] I. Mayergozy, Mathematical modeling of hysteresis and their application, Elsevier, Amsterdam, 2003.

[26] M.A. Krasnosel'skii, A.V. Pokrovskii, Systems with Hysteresis, Springer-Verlag, New York, 1989 (translated from Russian by M. Niezgodka).

[27] L. Liu, K.K. Tan, S.N. Huang, T.H. Lee, Identification and control of linear dynamics with input Preisach hysteresis, in: Proceedings of American Control Conference, Baltimore, USA, July 2010, pp. 4301-4306.

[28] L. Liu, K.K. Tan, S.L. Chen, C.S. Teo, L.H. Lee, Identification of coupled hysteresis, creep, electric and vibration dynamics in piezoelectric actuators for high bandwidth precision motion control, in: 12th International Conference of the European Society for Precision Engineering and Nanotechnology, Stockholm, Sweden, 2012.

[29] G.F. Franklin, J.D. Powell, M.L. Workman, Digital control of dynamic systems, third ed., Addition-Wesley, Menlo Park, 1997.

[30] J. Doyle, K. Glover, P. Khargonekar, B. Francis, State-space solutions to standard H_2 and H_∞ control problems, IEEE Trans. Autom. Control 34 (8) (1989) 831-847.

6

CASE STUDY OF A PIEZOELECTRIC STEERING PLATFORM

Abstract

The measurement and control strategy for a piezoelectric-based platform using strain gauge sensors and a robust composite controller is investigated in this chapter. First, we construct the experimental setup using a piezoelectric-based platform, strain gauge sensors, an AD5435 platform, and two voltage amplifiers. Then, we present the measurement strategy to measure accurately the tip/tilt angles on the order of subradians. Also, we present the comprehensive design of a composite control strategy to enhance the tracking accuracy with a novel driving principle. Finally, an experiment is presented to validate the measurement and control

Modeling and Precision Control of Systems with Hysteresis
http://dx.doi.org/10.1016/B978-0-12-803528-3.00006-9

strategy. The experimental results demonstrate that the proposed measurement and control strategy provides accurate angle motion with a root mean square error of 0.21 rad that is approximately equal to the noise level.

Keywords: Piezoelectric-based platform, Ultraprecision angle motion, Robust composite control

6.1 Introduction

Ultraprecision angle motion and positioning have significant requirements in free-space laser communication, space telescopes, staring cameras, and some other space optical instruments [1–4]. For instance, the angle-pointing accuracy of transmitters and receivers should be kept on the order of microradians to achieve free-space laser communication among satellites. Until recently, accurate angle motion and positioning in space applications were not solved well. To provide angle motion on the order of microradians, the measurement and control strategy for a piezoelectric-based platform with two degrees of freedom is investigated in this chapter by use of strain gauge sensors, which have been widely used for displacement measurement in the nanometer range. Intelligent actuators, which are commonly used to provide precision motion [5–7], replace classical electric motors in this chapter. Various intelligent actuators have been investigated for space applications [8–11]—for example, piezoelectric actuators, shape-memory-alloy actuators, and magnetostrictive actuators. Among these actuators, piezoelectric actuators are increasingly used as core components because of their properties, including high accuracy, high speed, and small size. Four piezoelectric actuators are used in this chapter to achieve accurate angle motion. A novel driving principle is used for the piezoelectric-based platform. Only one DC voltage and two varying voltages are used to drive four piezoelectric stacks. Additionally, high-bandwidth angle measurement on the order of subradians is necessary in applications of free-space optics. For example, angle acquisition and tracking with a bandwidth of hundred hertz is required in the acquisition, tracking, and pointing system of space laser communication. Generally, the tip and tilt (also named pitch and yaw) angles are measured and determined by attitude sensors (e.g., sun sensors, star sensors, integrated gyros, and fiber-optic gyros), but the accuracy is generally worse than 10 rad and the measurement bandwidth is generally less than 5 Hz. To measure tip and tilt angles on the order of subradians, strain gauge sensors are used in this chapter to determine the angles with high accuracy and high bandwidth through measurement of the length changes of piezoelectric actuators. Then, the tip and tilt angles are computed according to the length changes of the piezoelectric actuators. Further, electric bridges can be chosen for better stability and resolution [12]. To achieve fast angle motion on the order of subradians, closed-loop control is also necessary because of the linear dynamics and hysteresis effects in piezoelectric actuators [13–15]. It is known that the open-loop error of piezoelectric actuators can be as large as 10-15% of their travel span even at low frequencies (i.e., less than 0.1 Hz). Some compensation approaches for the hysteretic dynamics of piezoelectric systems have been proposed [16–19]. In this chapter, a comprehensive composite control strategy consisting of the proposed robust H feedback controller and a derivative feedforward controller is designed to compensate the linear dynamics and hysteresis effects efficiently. Instead of complex hysteretic dynamics, linear dynamics is used to represent the piezoelectric-based platform. The composite controller is designed from an application perspective. Compared with most other H_∞ work, the trade-off among the control bandwidth, measurement noise, and control limitation is quantitatively considered in our work. Moreover, a real-time physical simulation platform (i.e., AD5435) is used to provide real-time closed-loop control. The tracking error of the proposed robust composite control method approaches the measurement noise level. This chapter is organized as follows. First, the experiment setup and working principle are presented in Section 6.2. Next, the measurement scheme for the tip/tilt platform is proposed in Section 6.3. A piezoelectric-based platform, an AD5435 real-time platform, two amplifiers, and strain gauge sensors are used. Then, the composite control system for the piezoelectric-based platform is presented in Section 6.4. Finally, the experimental results are presented and discussed.

6.2   System Description

6.2.1   Experimental Setup

The experimental setup, the measurement strategy, and the angle calibration are presented in this section. The constructed piezoelectric platform system consists of a piezoelectric-based platform, strain gauge sensors, voltage amplifiers, an AD5435 platform, and signal conditioning, as shown in Figure 6.1. The AD5435 platform includes an AD5435 card, a digital-to-analog (DA) card (AD5430-02BK, eight channels, 16 bits), and an analog-to-digital (AD) card (AD5430-01, 16 channels, 16 bits). In addition, voltage amplifiers, strain gauge sensors, and signal conditioning are

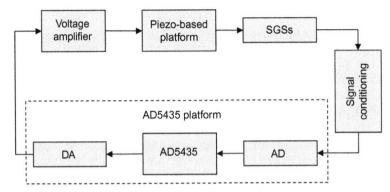

Figure 6.1 Experimental components for precision angle motion.

installed in a metallic box. The bandwidth of the strain gauge sensors in this chapter can be up to 3 kHz, and the resolution is better than 10 nm (root mean square, RMS). The equivalent angle resolution is 0.2 rad (RMS).

The experimental setup is shown in Figure 6.2. The piezoelectric-based platform system for experiments is constructed to achieve precision angle motion on the order of subradians.

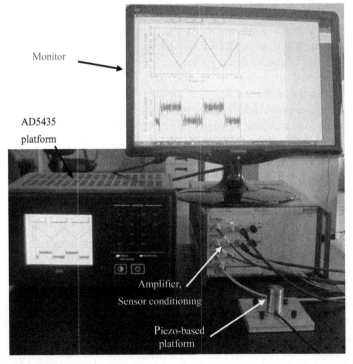

Figure 6.2 Experimental setup.

6.2.2 Working Principle

The working principle of the proposed piezoelectric-based platform is different from that of other piezoelectric applications. First, a preloading spring is used to maintain the structure stability. Then, a constant voltage and two varying inputs are used to drive four independent piezoelectric stacks. The working principle of the piezoelectric-based platform along the X-axis is shown in Figure 6.3, where r_1 and r_3 are the tilt angle, and the translational displacements of lead zirconate titanate (PZT) I and PZT III, respectively. A constant voltage +100 V is added to PZT I. Thus, the input voltage of channel 1 can be used to drive both PZT I and PZT III to produce the tilt angle. To further introduce the working principle, Figure 6.4 shows the working principle of the piezoelectric-based platform along both the X- and Y-axis. Channels 1 and 2 can be used to drive the four PZT stacks because a constant voltage of +100 V is preloaded onto two of the PZT stacks.

6.2.3 Measurement Strategy

In this section, we present the measurement strategy to obtain angle measurements with high accuracy on the order of subradians. Instead of direct measurements of the angle motion, the positions of the PZT stacks are measured first. Then the angle motion is computed. To evaluate the accuracy of the measurement strategy, the angle calibration is also presented.

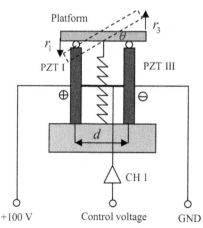

Figure 6.3 Working principle of the piezoelectric-based platform along the X-axis. CH, channel; GND, ground.

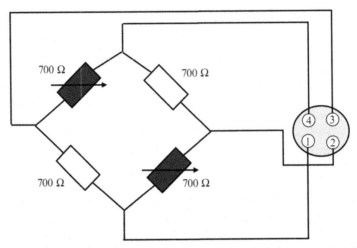

Figure 6.4 Measuring principle of the piezoelectric-based platform along both the X- and the Y-axis.

6.2.4 Measuring Principle of Strain Gauge Sensors

Strain gauge sensors are widely applied to measure the piezoelectric displacement. In this chapter, one piezoelectric actuator (PZT stack) is used to illustrate the measuring principle of strain gauge sensors. Two resistive films are bonded to the PZT stack, as shown in Figure 6.5. The length change of the piezoelectric stack alters the resistance of the strain gauge. The change in resistance

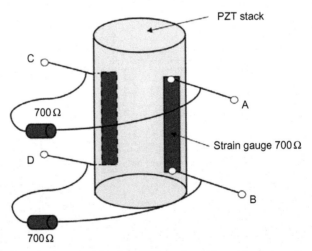

Figure 6.5 Measuring principle of the PZT stack position with strain gauge sensors.

is proportional to the displacement of the PZT stack. Moreover, a Wheatstone electric bridge, which consists of two strain gauges on the PZT stacks and two resistors on the PZT housing, is used to obtain better stability and resolution.

The strain gauges are used to measure the mechanical length change, because the length change results in a change in resistance and the generation of a voltage signal. More details of electric bridges have been presented by Huang et al. [12].

Figure 6.4 shows a sketch of the resistance bridge. Four resistors are used. Two of them are active and bonded to the PZT stack, and the other two are bonded to the housing. When the piezoelectric length changes, a stress is applied to the bonded strain gauge. Then, the Wheatstone bridge becomes unbalanced and there is a change in resistance. So a voltage signal is generated and it is proportional to the length change. With the conditioning electronics, the voltage is amplified to 0-10 V.

6.2.5 Measuring Principle for Tip/Tilt Angles

The tip/tilt angles of the piezoelectric-based platform are indirectly measured by comparison of the length changes of the PZT stacks. First, the strain gauge sensor gives the length change of the PZT stack. The relationship among θ, r_1, and r_3 in Figure 6.3 can be represented as

$$\tan \theta = \frac{r_1 + r_3}{d}. \tag{6.1}$$

The travel span of the piezoelectric stacks in this chapter is 15 μm. The maximum tip/tilt angles are 0.002 rad. Thus, $\tan \theta \approx \theta$. So θ can be represented by the following equation:

$$\theta = \frac{r_1 + r_3}{d}. \tag{6.2}$$

Finally, there is a linear relationship between the tilt angle and the piezoelectric displacements, and it will be validated in the next section.

6.2.6 Angle Calibration

The angle calibration is presented to validate the measurement strategy. For simplification, only the tilt angle calibration is presented. A ZYGO ZMI-2000 interferometer is used as the measurement setup to calibrate angles. The calibration temperature is 20.8 °C and the calibration humidity is 39%. Figure 6.6 shows the relationship between the sensor output voltage and the angle dis-

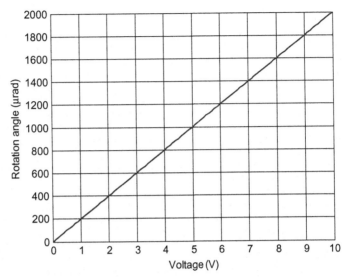

Figure 6.6 Relationship between the sensor output voltage and its angle displacement.

placement. It can be seen that there is a quasi-linear relationship between the sensor output voltage and the angle displacement.

Furthermore, Figure 6.7 shows the nonlinearity of the angle displacement measurement with respect to the sensor output voltage. The nonlinearity is represented by the relative error (%).

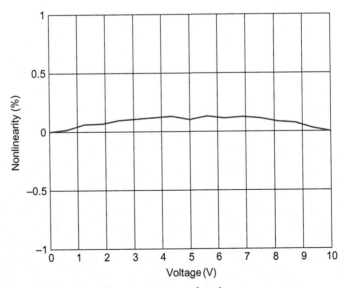

Figure 6.7 Nonlinearity of the measurement of strain gauge sensors.

It can be seen that the nonlinearity of the measurement is less than 0.1%. This demonstrates that the measurement strategy in this chapter is accurate and effective to measure the angle motion of the piezoelectric-based platform.

6.3 Control System Design

In this section, we present the comprehensive design of the composite control of the piezoelectric-based platform to achieve precision angle motion. Only the tilt motion (i.e., along the X-axis) control is investigated. The coupling effect of tilt and tip motion of the piezoelectric-based platform is represented by an input uncertainty. First, the linear dynamics of the piezoelectric-based platform is identified. Then, the robust H_∞ feedback controller is designed. Sequentially, the derivative feedforward is incorporated to enhance the tracking performance. Finally, the proposed composite controller is applied to the piezoelectric-based platform to demonstrate the tracking performance.

6.3.1 Modeling and Identification

Modeling and identification are necessary for the design of modern robust control systems. Various modeling and identification methods for piezoelectric systems have been investigated [20–23]. To design a robust optimal H_∞ controller, time-invariant linear dynamics is used to represent the main dynamics of the PZT stacks. Also, the unmodeled dynamics is represented by an input uncertainty. The time-invariant linear dynamics of the piezoelectric-based platform is identified by use of a square-wave input (only the X-axis is investigated). By use of the autoregressive moving average model with external input method, the linear dynamics of the piezoelectric stack is identified and represented by the following transfer function:

$$G = 422 \frac{(s^2+306.7s+5.022\times10^6)(s^2-5.605\times10^4s+7.938\times10^8)(s^2+2358s+1.052\times10^8)}{(s+1996)(s^2+1021s+5.171\times10^6)(s^2+5816s+6.632\times10^7)(s^2+6364s+3.156\times10^8)},$$

$$(6.3)$$

where G represents the time-invariant linear dynamics of the piezoelectric-based platform system and s denotes a differentiation operator.

6.3.2 Design of Robust Optimal Control

Multiobjective robust H_∞ optimal control is investigated in this section to achieve precision angle motion in the face of modeling uncertainty and measurement noise. Weighting func-

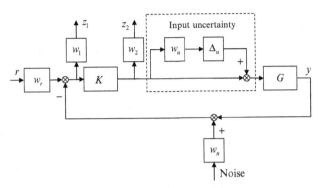

Figure 6.8 Design sketch of the robust H_∞ controller.

tions and a loop-shaping technique are used to design the robust H_∞ controller. Figure 6.8 shows the design sketch of the robust H_∞ controller, in which K and G represent the robust H_∞ controller and the linear dynamics of the PZT stacks, respectively, and w_1 is the performance weighting function. To enhance the disturbance suppression at low frequencies and limit the feedback gain at high frequencies simultaneously, an integral action is added to w_1, and w_2 is the control weighting function which limits the control gain at high frequencies. To guarantee robust stability at high frequencies, the control signal is limited to less than 2000 Hz. w_r and w_n represent the reference and noise weighting functions, respectively. In the experiment, the measurement noise level at frequencies higher than 500 Hz is 100 times the noise level at frequencies lower than 5 Hz. The weighting function w_u denotes the modeling uncertainty mainly produced by the hysteresis non-linearity of the PZT stacks, and Δu is a unit complex dynamic uncertainty with $\| \Delta u \|$. After the hysteresis loop has bene measured in experiments, the weighting function w_u is set to 5%.

After several trials, the weighting functions w_r, w_u, w_1, w_2, and w_n are designed and are represented by the following equations:

$$w_r = 1, \quad w_u = 0.05, \tag{6.4}$$

$$w_1 = \frac{3000}{s + 0.0001}, \tag{6.5}$$

$$w_2 = 0.5 \left(\frac{s + 400\pi}{s + 4000\pi} \right)^2, \tag{6.6}$$

$$w_n = 0.01 \frac{s + 10\pi}{s + 1000\pi}. \tag{6.7}$$

The *D-K* iteration approach with structured singular values is used to solve the H optimal controller [24]. After 10 iteration steps, the final structure singular value μ is 0.95, which is less than 1. The small-gain theorem is satisfied, and the H_∞ feedback control is stable. Moreover, its order is reduced to 8, such that it is easier to implement the H_∞ controller in an AD5435 card. The reduced H_∞ optimal controller is represented by the following equation:

$$K = 2.04 \frac{(s+2215)(s^2+1119s+4.872\times10^6)(s^2+4579s+8.388\times10^7)}{(s+0.0001)(s^2+375.9s+4.943\times10^6)(s^2+3079s+9.535\times10^7)}$$
$$\cdot \frac{(s^2+6893s+4.233\times10^9)}{(s^2+6359s+3.958\times10^9)}. \tag{6.8}$$

6.3.3 Composite Control Design

To further enhance the tracking performance of the proposed robust H_∞ feedback control, a derivative feedforward of the reference signal is added to the tacking error. The composite control is thus designed, as shown in Figure 6.9, where *D* and *K* represent the derivative feedforward controller and the robust H_∞ feedback controller, respectively, and *r* and *y* represent the reference angle and piezoelectric angle, respectively. In the experiment, *D* is determined by several trials.

Finally, *D* is represented as follows:

$$D = 0.0004s, \tag{6.9}$$

where *s* represents the derivative operator.

In this chapter, the derivative feedforward *D* is used to compensate the phase lag within the feedback bandwidth. By transformation, the proposed composite controller can be

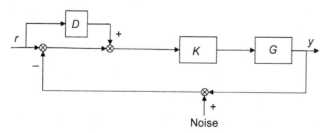

Figure 6.9 Composite control sketch.

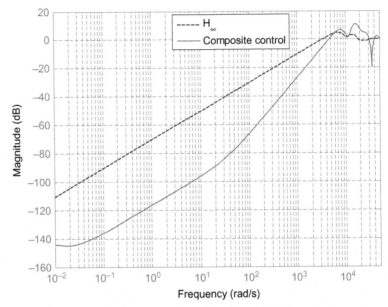

Figure 6.10 Bode diagram of the sensitivity function of the H_∞ controller and composite controller.

regarded as a special proportional-derivative (PD)-H_∞ composite controller.

To represent the tracking performance in the frequency domain, the Bode diagram of the sensitivity functions of the H_∞ controller and the composite controller is shown in Figure 6.10. It can be seen that the proposed composite controller greatly improves the tracking performance at frequencies lower than 3500 rad/s. At frequencies higher than 5000 rad/s, the tracking performance of the proposed composite controller does not hold. Alternatively, robust stability is guaranteed by the robust H_∞ feedback controller at frequencies higher than 5000 rad/s.

6.3.4 Simulation Result for the Proposed Composite Controller

In this section, a simulation study with a triangular wave with an amplitude of 100 μrad and a frequency of 10 Hz is presented. Figure 6.11 shows the tracking errors of the proposed composite controller and the H_∞ controller. The tracking error of the proposed composite control is less than 36% of that of the H_∞

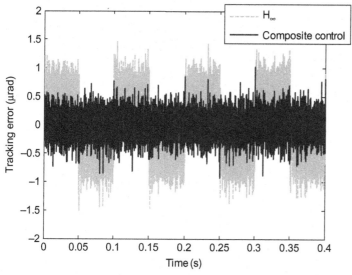

Figure 6.11 Tracking errors in the simulation study.

feedback controller. The tracking error RMS of the proposed composite control is 0.205 µrad, which is almost equal to the electrical noise. The simulation result indicates that the proposed composite controller has satisfactory tracking performance.

6.4 Experimental Results and Discussion

6.4.1 Experimental Results

In this section, the experimental studies are investigated to demonstrate the measurement and control strategy. Both triangular and sinusoidal waves are selected as the reference signal. First, a triangular wave of 10 Hz is chosen as the reference signal r. Figure 6.12 shows the triangular wave tracking performance of the proposed composite control. It can be seen that the piezoelectric-based platform tracks the reference angle with satisfactory performance.

The triangular wave tracking error of the proposed composite controller is shown in Figure 6.13. Compared with the simulation result in Figure 6.11, the performance degradation of the H_∞ controller and the composite controller is negligible. The tracking error of the proposed composite controller is less than 0.8 µrad, with an RMS error of 0.21 µrad. The relative tracking error of the proposed composite controller is 0.36% of the reference signal RMS error. The following equation is used to represent the relative tracking error:

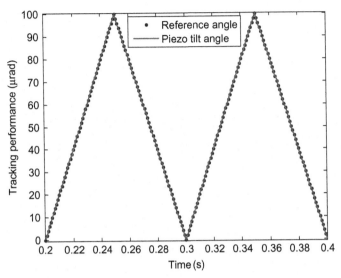

Figure 6.12 Tracking performance for a triangular wave.

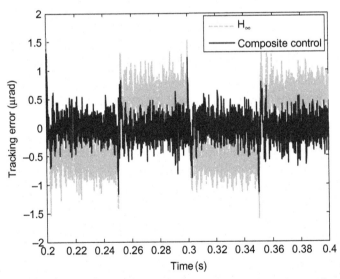

Figure 6.13 Tracking error for a triangular wave in the piezoelectric experiments.

$$e_{\text{RMS}} = \left(\frac{\sqrt{\frac{1}{n} \sum_{i=1}^{n} (y_{\text{r}}(i) - y(i))}}{\sqrt{\frac{1}{n} \sum_{i=1}^{n} y_{\text{r}}(i)}} \right), \qquad (6.10)$$

where n is the sampling number, y_{r} is the reference angle, and y is the output angle.

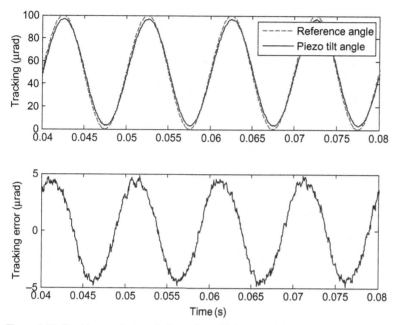

Figure 6.14 Tracking performance for a sinusoidal wave of 100 Hz under composite control.

To further demonstrate the high-frequency tracking performance of the proposed composite controller, the tracking of a sinusoidal wave of 100 Hz is given in Figure 6.14. It can be seen that satisfactory performance is obtained. The relative tracking error of the proposed composite controller is 5.7%.

6.4.2   Discussion

The experimental results reveal that precision angle motion of the piezoelectric-based platform can be achieved by use of the proposed composite controller. In addition, for angle motion and positioning with high accuracy, linearity, and repeatability on the order of subradians, closed-loop operation is necessary for the piezoelectric-based platform although the piezoelectric-based platform is commonly regarded as a precision system. Thus, a robust composite controller is designed in a closed loop to compensate the linear dynamics and hysteresis effects. It can be seen from the results that the tracking accuracy for the triangular wave of 10 Hz approaches the level of measurement noise. For comparison, the experimental results presented here are compared with those in [22], where the complex hysteresis effect was modeled and compensated. In [22], the relative tracking error at 0.01 Hz is

4.37%, obtained with a composite controller consisting of inverse-Preisach hysteresis and PD/lead-lag feedback. In this chapter, the relative tracking error at 10 Hz obtained with the proposed composite controller is as small as 0.37%. Furthermore, the reference signal is faster than that in [22], where the maximum testing frequency was 0.1 Hz, but the maximum reference frequency in this chapter is 100 Hz. This indicates the proposed composite controller in this paper results in more accurate tracking at higher bandwidth.

6.5 Conclusions

Precision angle motion and positioning in space optics are increasingly demanded. A piezoelectric-based platform system was constructed and the measurement and control strategy was presented to resolve this problem. The measurement strategy was investigated to provide effective measurement with high bandwidth and accuracy. A comprehensive design of the composite control strategy was developed. The effectiveness of precision angle motion under the proposed composite controller was validated through the experimental results. The results demonstrate that the measurement and control strategy presented in this chapter will have a significant benefit for future applications in space optics.

References

[1] H. Hemmati, A. Biswas, I. Djordjevic, Deep-space optical communications: future perspectives and applications, Proc. IEEE 99 (2011) 2020-2039.

[2] W. Roberts, R. Jeffrey, Compact deep-space optical communications transceiver, in: Proceedings of SPIE Conference on Free-Space Laser Communication Technologies XXI, vol. 7199, 2009, Paper no. 71990L.

[3] P. Brugarolas, J. Alexander, J. Trauger, et al., Access Pointing Control System, in: Proceedings of SPIE Conference on Space Telescope and Instrumentation, vol. 7731, 2010, Paper no. 77314V.

[4] S. Kendrick, J. Stober, I. Gravseth, Pointing and image stability for spaceborne sensors-from comet impactors to observations of extrasolar planets, in: Proceedings of SPIE Conference on Space Telescope and Instrumentation, vol. 6265, 2006, Paper no. 62652V.

[5] L. Ladani, S. Nelson, A Novel piezo-actuator-sensor micromachine for mechanical characterization of micro-specimens, Micromachines 1 (2010) 129-152.

[6] Y. Yang, B. Divsholi, C. Soh, A reusable PZT transducer for monitoring initial hydration and structural health of concrete, Sensors 10 (2010) 5193-5208.

[7] S. Jakiela, J. Zaslona, J. Michalski, Square wave driver for piezoceramic actuators, Actuators 1 (2012) 12-20.

[8] H. Anderson, J. Sater, Smart Structures Product Implementation Award: a Review of the First Ten Years, in: Proceedings of SPIE Conference on

Industrial and Commercial Applications of Smart Structures Technologies, vol. 6527, 2007, Paper no. 652702.

[9] R. Cobb, J. Sullivan, A. Das, Vibration isolation and suppression system for precision payloads in space, Smart Mater. Struct. 43 (1999) 798-812.

[10] G. Neat, J. Melody, J. Lurie, Vibration Attenuation Approach for Spaceborne Optical Interferometers, IEEE Trans. Control Syst. Technol. 6 (1998) 689-700.

[11] A. Bronowicki, J. Innis, A family of full spacecraft-to-payload isolators, Tech. Rev. J. 13 (2005) 21-41.

[12] H. Huang, H. Zhao, Z. Yang, Z. Fan, S. Wan, C. Shi, Z. Ma, Design and analysis of a compact precision positioning platform integrating strain gauges and the piezoactuator, Sensors 12 (2012) 9697-9710.

[13] S. Moheimani, Accurate fast nanopositioning with piezoelectric tube scanners: emerging trends and future challenges, Rev. Sci. Instr. 79 (2008) Paper no 0711011.

[14] T. Ando, T. Uchihashi, T. Fukuma, High-speed atomic force microscopy for nano-visualization of dynamic biomolecular processes, in: Prog. Surface Sci. 83 (2008) 337-437.

[15] H. Xie, M. Rakotondrabe, S. Rgnier, Characterizing piezoscanner hysteresis and creep using optical levers and a reference nanopositioning stage, Rev. Sci. Instr. 80 (2009) Paper no. 046102.

[16] D. Huang, J. Xu, T. Huynh, High performance tracking of piezoelectric positioning stage using current-cycle iterative learning control with gain scheduling, IEEE Trans. Indust. Electron. (2013) http://dx.doi.org/10.1109/TIE.

[17] T. Yan, X. Xu, J. Han, R. Lin, B. Ju, Q. Li, Optimization of sensing and feedback control for vibration/flutter of rotating disk by PZT actuators via air coupled pressure, Sensors 11 (2011) 3094-3116.

[18] G. Song, J. Zhao, X. Zhou, Tracking control of a piezoceramic actuator with hysteresis compensation using inverse Preisach model, IEEE/ASME Trans. Mechatron. 10 (2005) 198-209.

[19] S. Huang, K. Tan, T. Lee, Adaptive Sliding-Mode Control of Piezoelectric Actuators, IEEE Trans. Ind. Electron. 56 (2009) 3514-3522.

[20] T. Jin, A. Takita, M. Djamal, W. Hou, H. Jia, Y. Fujii, A method for evaluating the electro-mechanical characteristics of piezoelectric actuators during motion, Sensors 12 (2012) 11559-1570.

[21] A. Putra, K. Ridge, S. Huang, K. Tan, S. Panda, Design, modeling and control of piezoelectric actuators for intracytoplasmic sperm injection, IEEE Trans. Control Syst. Technol. 15 (2007) 879-890.

[22] G. Huang, F. Song, X. Wang, Quantitative modeling of coupled piezo-elastodynamic behavior of piezoelectric actuators bonded to an elastic medium for structural health monitoring: a review, Sensors 10 (2010) 3681-3702.

[23] Y. Cao, X. Chen, State space system identification of 3-degree-of-freedom (DOF) piezo-actuator-driven stages with unknown configuration, Actuators 2 (2013) 1-18.

[24] S. Skogestad, I. Postlethwaite, Multivariable Feedback Control: Design and Analysis, Wiley, New York, 2005, 373-377.

7

CASE STUDY OF ACTIVE VIBRATION ISOLATION

Abstract

A case study of active vibration isolation is presented in this chapter. Multiobjective robust H_∞ and μ synthesis strategies, based on singular values and structured singular values, respectively, are presented, and simultaneously satisfy the low-frequency pointing and high-frequency disturbance rejection requirements and take account of the model uncertainty, parametric uncertainty, and sensor noise. Then, by a robust stability test, it is shown that the two controllers are robust to the uncertainties; the robust stability margin of the H_∞ controller is less than that of the μ controller, but the order of the μ controller is higher than that of the H_∞ controller, so balanced controller reduction is provided. Additionally, the μ controller is compared with a proportional-integral controller. The time domain simulation of the μ controller indicates that the two robust control strategies are effective in

Modeling and Precision Control of Systems with Hysteresis
http://dx.doi.org/10.1016/B978-0-12-803528-3.00007-0

keeping the pointing command and isolating the sinusoidal and stochastic disturbances.

Keywords: Active vibration isolation, Robust H_∞, μ synthesis, Loop shaping

7.1 Introduction

The requirements for precision pointing accuracy and extreme stability of modern optical spacecrafts are becoming stricter. The James Webb Space Telescope, terrestrial planet finders, space-based laser weapons, space-based interferometers and deep-space laser communication are examples in which sub-microradian pointing and nanometer level of motion stability are required [1–6]. On the other hand, space systems contain various vibration sources. A satellite may host multiple motion instruments, some of which may use reaction wheels, cryogenic coolers, control moment gyroscopes, solar array drives, stepper motors, and other motion devices. These devices will transmit vibrations over a broad range of frequencies. As a result, vibration suppression is increasingly needed for future space telescopes, intersatellite laser communication, and other space-sensitive payloads which require ultraquiet dynamics environments.

First, passive isolation is a reliable, low-cost solution that is effective in attenuating high-frequency vibrations, but it is, in general, not suitable for low-frequency vibration isolation, and especially, passive isolation cannot provide a good trade-off between resonant peak and high-frequency attenuation and between keeping the pointing command and disturbance rejection [6]. Alternatively, active vibration control can overcome these limitations.

To achieve vibration isolation and precision pointing with multiple degrees of freedom, the Stewart platform (or hexapod), especially the cubic one, has become one of the most popular approaches [7–13], as shown in Figure 7.1. The cubic hexapod simplifies the control topologies and allows the decoupled controller designs to be identical for each strut [12, 13]. To eliminate the microdynamics (friction and backlash), flexure joints are generally used [9, 13].

The Jet Propulsion Laboratory, the Air Force Research Laboratory, the Naval Postgraduate School, the University of Washington, the Free University of Brussels, the University of Wyoming, CSA Engineering, other organizations are very active in this field [2, 8, 10]. Classic control, adaptive control, linear-quadratic-Gaussian control, neural control, simple robust control, and other control approaches have been studied [10, 11, 14–18]. In this chapter, the dynamics and controllers are designed for the struts of Stewart platforms to suppress the overshoot at the

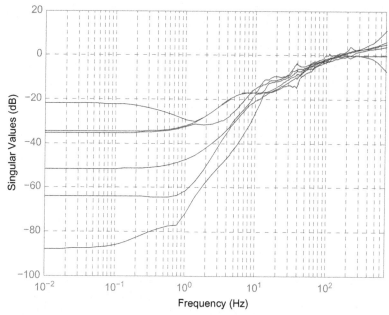

Figure 7.1 Singular value curves for the transfer matrix of SUITE.

resonant frequency, while keeping the pointing signal transmiting with weak dropping. Finally, frequency-domain and time-domain simulation studies are presented.

7.2   Multiobjective Robust Active Vibration Control for Flexure Jointed Struts

A Stewart platform-based hybrid isolator with six hybrid struts is an effective system for active/passive vibration isolation in the 5-250 Hz band. With use of an identification transfer matrix of the Stewart platform, the coupling analysis of six channels is presented. A dynamic model is derived, and the rigid mode is removed to keep the signal of pointing control.

7.3   Coupling Analysis and Dynamics Model

Differently from the gains of a single-input single-output (SISO) system, the gains of a multiple-input multiple-output (MIMO) system, known as the singular values or the directional gains, depend on the direction of the input vector. The transfer function of the Stewart platform is a 6×6 matrix. It behaves as a SISO system for a weak coupling MIMO system. The condition number, defined by (7.1), determines the coupling behavior of a MIMO system.

$$k(\mathbf{G}(jw)) = \frac{\overline{\sigma}(\mathbf{G}(jw))}{\underline{\sigma}(\mathbf{G}(jw))}, \qquad (7.1)$$

where $\mathbf{G}(jw)$ is the transfer matrix, $\overline{\sigma}$ is the largest singular value of the matrix, and $\underline{\sigma}$ is the smallest one.

Referring to Joshi [18] and Joshi and Kim [19], the identification transfer matrix of SUITE (a cubic lead zirconate titanate (PZT) Stewart platform built by the Air Force Research Laboratory) is used to analyze the coupling, and the six singular values are plotted in Figure 7.1.

Figure 7.1 shows the six singular values, which represent the six principal input directions. It can be seen that the singular value band is from 10 to 500 Hz and is quite narrow, and the corresponding condition numbers are approximately equal to 1, which means that the system will be insensitive to the direction of the input vector.

Furthermore, the diagonal entries are extracted to form a new system G' without coupling. Figure 7.2 shows their singular value curves. It can be seen that the two systems behave similarly within the 10-500 Hz frequency range. So the effects of the nondiagonal entries can be regarded as the parasitic stiffness and damping of a single strut, and the MIMO system is degraded to six SISO systems.

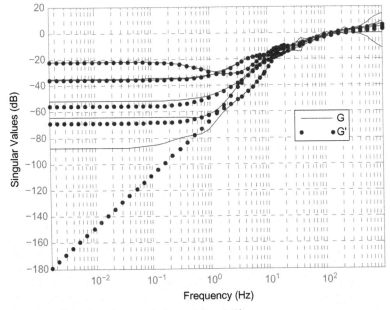

Figure 7.2 Singular value comparison of G and G'.

7.3.1   Nonhysteretic Dynamics

According to the coupling analysis for the six struts of the cubic Stewart platform, the hexapod is decoupled into six single-axis systems. Figure 7.3 shows the model sketch of a single strut, and the measuring signal is the force of the strut on the side of the payload. But the forces due to the parasitic stiffness and damping, which represent the coupling between six struts, are not involved in the measured output [14].

In Figure 7.3, the mass of the base $m_b = 200$ kg, the mass of the strut $m_s = 0.254$ kg, the mass of the payload $m_p = 20$ kg, the parasitic stiffness $k_p = 760$ N/m, the parasitic damping $c_p = 2$ kg/s, the axis stiffness of flexure joints $k_2 = 800$ kN/m, the damping of flexure joints $c_2 = 100$ kg/s, the stiffness of the actuator $k_1 = 80$ MN/m, the damping of the actuator $c_1 = 100$ kg/s, u is the output of the actuator, and r is the attitude control signal under disturbance. The model of a signal strut shown in Figure 7.3 can be expressed by

$$\mathbf{M\ddot{q}} + \mathbf{D\dot{q}} + \mathbf{Kq} = \mathbf{f},$$
$$\mathbf{y} = \mathbf{C_q q} + \mathbf{C_v \dot{q}},\tag{7.2}$$

where $\mathbf{q}$, $\mathbf{M}$, $\mathbf{f}$, $\mathbf{C_q}$, $\mathbf{C_v}$, $\mathbf{D}$, and $\mathbf{K}$ are written as

$$\mathbf{q} = (x_b, x_s, x_p)^T,$$
$$\mathbf{M} = \text{diag}([m_b, m_s, m_p]),$$
$$\mathbf{f} = \begin{pmatrix} 1 & 0 & 0 \\ 1 & -1 & 0 \end{pmatrix}^T \cdot \begin{bmatrix} r \\ u \end{bmatrix},$$
$$\mathbf{C_q} = [\ 0 \quad -k_2 \quad k_2\ ],$$

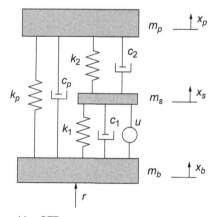

Figure 7.3 The strut with a PZT actuator.

$$\mathbf{C_v} = [\ 0 \quad -c_2 \quad c_2\ ],$$

$$\mathbf{D} = \begin{pmatrix} c_1 + c_2 & -c_1 & -c_p \\ -c_1 & c_1 + c_2 & -c_2 \\ -c_p & -c_2 & c_p + c_2 \end{pmatrix},$$

$$\mathbf{K} = \begin{pmatrix} k_1 + k_p & -k1 & -k_p \\ -k_1 & k_1 + k_2 & -k_2 \\ -k_p & -k_2 & k_2 + k_p \end{pmatrix}.$$

By introducing $\mathbf{K}$ and $\mathbf{M}$ in the equation $\det(\mathbf{K} - \omega^2\mathbf{M}) = 0$, we can obtain the modal frequency matrix $\Omega = \mathrm{diag}([w_1, w_2, w_3])$ as

$$\Omega = \begin{pmatrix} 0 & 0 & 0 \\ 0 & 207.8 & 0 \\ 0 & 0 & 11670 \end{pmatrix}. \tag{7.3}$$

If we substitute Ω into $(\mathbf{K} - \Omega^2\mathbf{M})\Phi = 0$ [18], the modal matrix is given by

$$\Phi = \begin{pmatrix} -0.5774 & -0.0015 & -0.0991 \\ -0.5774 & -0.9999 & -0.0777 \\ -0.5774 & 0.0003 & 0.9920 \end{pmatrix}. \tag{7.4}$$

Introducing the modal matrix Φ into (7.2), we obtain the modal model as shown in (7.5):

$$\begin{cases} \mathbf{M}_m\ddot{\mathbf{q}}_m + \mathbf{D}_m\dot{\mathbf{q}}_m + \mathbf{K}_m\mathbf{q}_m = \Phi^T\mathbf{f}, \\ \mathbf{y} = \mathbf{C}_q\Phi\mathbf{q}_m + \mathbf{C}_v\Phi\dot{\mathbf{q}}_m, \end{cases} \tag{7.5}$$

where $\mathbf{q}_m = \Phi^{-1}\mathbf{q}$, $\mathbf{D}_m = \Phi^T\mathbf{D}\Phi$, $\mathbf{M}_m = \Phi^T\mathbf{M}\Phi$, and $\mathbf{K}_m = \Phi^T\mathbf{K}\Phi$.

It is seen that the first mode is a rigid mode, and the corresponding natural frequency is zero (in Ω). On the other hand, in the singular value plot for the model with parametric (stiffness and damping) uncertainty, as illustrated in Figure 7.4, a rigid mode can be found. The structures with rigid modes are unstable, but the rigid mode is the one that allows a controller to move the structures or track a command [20]. Thus, the rigid mode is removed when an active vibration isolation controller is designed. The singular value response for the dynamic model without a rigid mode is shown in Figure 7.5, where the parametric uncertainty is also included.

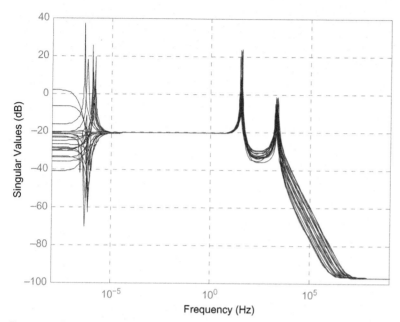

Figure 7.4 Singular value diagram of the proposed Stewart struts.

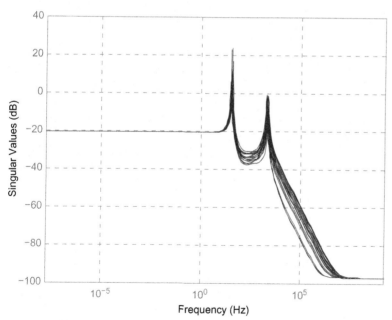

Figure 7.5 Singular value diagram of the proposed Stewart struts without the rigid mode.

7.4 Robust Synthesis Controller Design

The active vibration controller should be robust to the modeling uncertainty and parametric uncertainty because of the complex dynamics environment and modeling error. Robust H_∞ synthesis and μ synthesis are thus presented in this work, and the dynamic uncertainty is represented by the output uncertainty. The parameter uncertainty is also included in the robust stability and performance test.

7.4.1 Performance and System Interconnection

The performance objective of this work is proposed with the strict requirements of future precision spacecrafts—that is, the low-frequency pointing signals must be fully transmitted through the Stewart platform, but the high-frequency disturbances (both harmonic and broadband), which will disturb the precision instruments, should be isolated. The two strict requirements are reformed as follows:

REQ 1: For a low-frequency pointing command (0-5 Hz), the pointing actuating signal is attenuated less than 0.2 dB.

REQ 2: Disturbance suppression requirement. The overshoot at the resonant frequency is suppressed more than 25 dB.

This function is also known as active damping.

The structure of the closed-loop system is shown in Figure 7.6. G denotes the linear dynamics without the rigid mode, K denotes the controller to be designed, and the weighting functions are used to represent the magnitude, relative importance, and frequency content of inputs and outputs [21]. The performance weighting function W_1 denotes the relative significance of the performance requirements at difference frequencies.

The maximum peak of G is 23 dB, so the maximum of W_1 is set to more than 0 dB to satisfy requirement 1. The control weighting

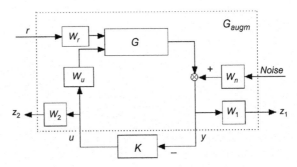

Figure 7.6 Closed-loop system structure in the robust design.

function W_2 is used to avoid the saturation of the PZT actuator and suppresses the high- and low-frequency gains. Generally, the maximum force of actuators is 400 N, and W_2 is set to more than -52 dB (1/400). The noise weighting function W_n is set to less than 0.3 N at low frequencies (below 300 Hz), and is set to 1 N at high frequencies(above 1000 Hz). W_r is the disturbance weighting function. The weighting functions are selected as follows:

$$W_1 = \frac{0.000445s^2 + 70s + 0.00022}{s^2 + 64.4s + 98.7}, \tag{7.6}$$

$$W_2 = \frac{0.3s^2 + 154.2s + 18850}{s^2 + 62830s + 314.2}, \tag{7.7}$$

$$W_n = \frac{s + 942.5}{s + 9425}. \tag{7.8}$$

The augmented plant G_{augm} is given by

$$\begin{cases} \dot{x} = \mathbf{A} + \mathbf{B}_1 w + \mathbf{B}_2 u, \\ z = \mathbf{C}_1 x + \mathbf{D}_{11} w + \mathbf{D}_{12} u, \\ y = \mathbf{C}_2 x + \mathbf{D}_{21} w + \mathbf{D}_{22} u, \end{cases} \tag{7.9}$$

where $z = [z_1, z_2]^T$ and $w = r$

The PZT stacks are high-precision actuators, but they are typically not highly linear; for the nonlinear factors, such as hysteresis, creep, and temperature effects, and at low frequencies the error can be 10-15% of the full scale in the open loop. And so the output uncertainty of the PZT stack is represented by W_u , as shown in Figure 7.7:

$$W_u = 1 + \frac{1}{10} \cdot \frac{2094s + 10^6}{12.6s + 10^6}. \tag{7.10}$$

7.4.2 Controller Design

The H_∞ synthesis is a mixed sensitivity H_∞ suboptimal control, based on the Doyle-Glover-Khargonekar-Francis method [22], and μ synthesis is based on iteration [21, 23–25]. The following criterion is used for H_∞ synthesis:

$$\left\| \begin{matrix} W_1(s)\mathbf{S}(s) \\ W_2(s)\mathbf{T}(s) \end{matrix} \right\|_{H_\infty} < \gamma, \tag{7.11}$$

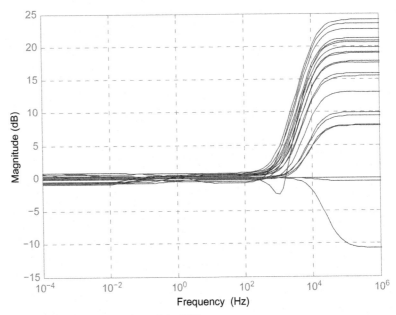

Figure 7.7 Output uncertainty of the PZT actuator.

where $\mathbf{S}(s)$ and $\mathbf{T}(s)$ are the sensitivity function and complementary sensitivity function, respectively.

The D-K iteration μ synthesis method is based on solving the following optimization problem, for a stabilizing controller $\mathbf{K}$ and a diagonal constant scaling matrix $\mathbf{D}$:

$$\mathbf{K}(s) = \arg \min_{K(s) \in \kappa(s)} \sup_{\omega \in \mathbb{R}} \inf_{\mathbf{D}(s) \in \mathbf{D}} \left\{ \mathbf{D}(s) F_l(\mathbf{P}, \mathbf{K}) \mathbf{D}^{-1}(s) \right\}, \qquad (7.12)$$

where $\mathbf{P}$ is the open-loop interconnected transfer function matrix of the system.

The D-K iteration procedure can be formulated as follows:

Step 1: Start with an initial guess for $\mathbf{D}$, usually set $\mathbf{D} = \mathbf{I}$.

Step 2: Fix D and solve the H_∞ suboptimal $K(s)$:

$$\mathbf{K}(s) = \arg \min_{K(s) \in \kappa s} \| F_l(\mathbf{P}\mathbf{K}) \|_{H_\infty}. \qquad (7.13)$$

Step 3: Fix $K_i(s)$ and solve the convex optimal problem for $\mathbf{D}^*$ at each frequency over a selected frequency range:

$$\mathbf{D}^*(J\omega) = \arg \min_{D \in \mathbf{D}} \overline{\sigma} \left(\mathbf{D}(s) F_l(\mathbf{P}, \mathbf{K}) \mathbf{D}^{-1}(s) \right). \qquad (7.14)$$

Step 4: Curve-fit $\mathbf{D}^*(j\omega)$ to get a stable, minimum phase $\mathbf{D}^*$, and compare $\mathbf{D}^*$ and $\mathbf{D}$. Stop if they close in magnitude, otherwise go to step 2 until the tolerance is achieved.

The H_∞ norm γ is solved to be 0.9932 and the 10th-order H_∞ controller is obtained. Alternatively, the structured singular value μ is found to be 0.993, and the 12th-order μ controller is obtained. The Bode magnitudes of the two controllers are given in Figure 7.8.

During the control synthesis process, the weighting functions W_1 and W_2 are adjusted repeatedly (a few trials are needed), and the final results are given by (7.11).

The closed-loop structure without performance weighting functions is shown in Figure 7.9, where G contains a rigid mode.

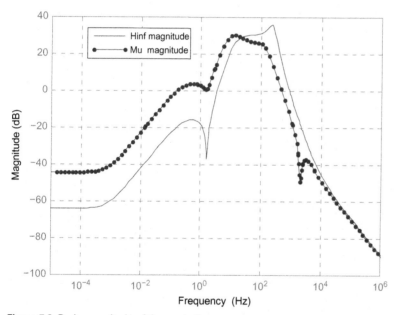

Figure 7.8 Bode magnitude of the controllers.

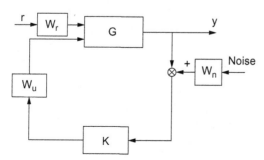

Figure 7.9 Closed-loop system structure for frequency responses.

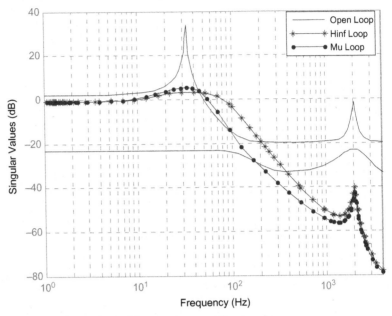

Figure 7.10 Comparison of the open loop and the closed loop.

The singular value plots for closed loops are shown in Figure 7.10, from which it can be seen that H_∞ and μ controllers isolate high-frequency disturbance and noise in the neighborhood of resonant frequencies. The disturbance and noise isolated by the H_∞ controller is more than 27 dB, and for the μ controller the value is 21 dB.

Figure 7.11 shows that low-frequency pointing fully transfers with attenuation less than 0.2 dB. The nominal performance for the H_∞ synthesis controller is better than that for the μ synthesis controller at resonance but worse at high frequencies.

7.4.3 Robust Stability Analysis

Robust stability is very important because of various uncertainties [21], and in this section we give the robust stability margins of the uncertain closed loop. By calculation, the robust stability margin for the H_∞ closed loop is 1.56, and the destabilizing frequency is 625.9 rad/s; the corresponding values are 6.29 and 346 rad/s for the μ closed loop. Their stability robustness margins greater than 1 mean that the uncertain system is stable for all values of its modeled uncertainty.

On the other hand, parametric uncertainty, which is a 30% change in stiffness and 80% change in damping, is considered with modeling uncertainty to test the robust stability and robust performance further. Figures 7.12 and 7.13 show singular value

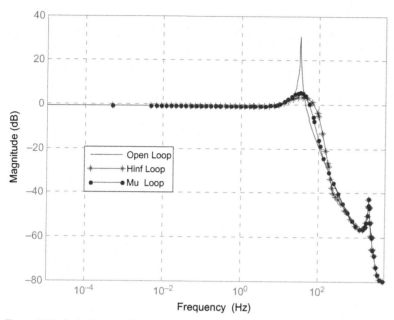

Figure 7.11 Bode diagram from r to y.

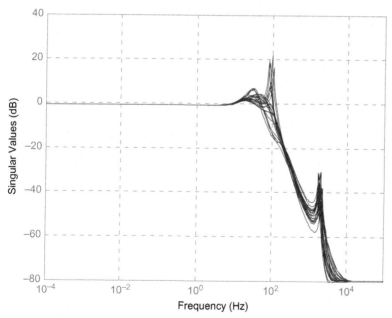

Figure 7.12 Singular value plot for the H_∞ closed loop.

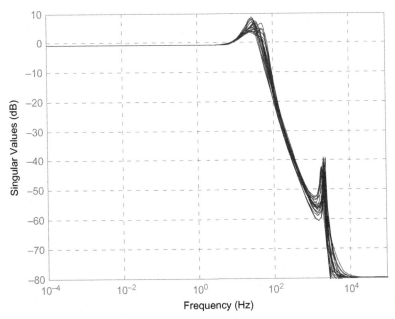

Figure 7.13 Singular value plot for the μ closed loop.

plots for the closed loop, from which it can be seen that robust stability and robust performance for the H_∞ closed loop are worse than for the μ closed loop in the presence of large uncertainty.

7.4.4   Controller Reduction

In this chapter, the full order of the H_∞ controller is 10 and the full order of the μ controller is 12. In real-time control, reduced-order control is easier to implement. Thus, square root balanced model truncation [26–28] is used to reduce the controller order. Figure 7.14 shows Bode diagrams of the sixth-order H_∞ controller and eighth-order μ controller as well as the original H_∞ and μ controllers.

The stability robustness margin is 1.56 for the reduced H_∞ closed loop and 6.3 for the μ closed loop, so the reduced controllers have robust stability.

7.5   Performance of Active Vibration Isolation

7.5.1   Results of Active Vibration Isolation

The frequency-domain simulation of the open-loop and closed-loop systems was presented in the previous section. In

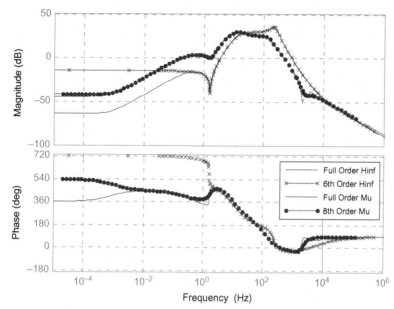

Figure 7.14 Bode diagrams of the full-order and reduced-order controllers.

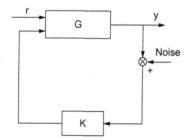

Figure 7.15 Nominal closed-loop system structure.

this section we give the corresponding transient response for a reduced-order μ controller and a proportional-integral (PI) controller, shown in (7.15), for the nominal closed-loop system structure shown in Figure 7.15 for time-domain responses.

$$K_{PI} = \frac{20}{s} + 32, \tag{7.15}$$

$$r_0 = 10 \sin t, \tag{7.16}$$

$$\text{dist} = 0.1 \sin(66\pi). \tag{7.17}$$

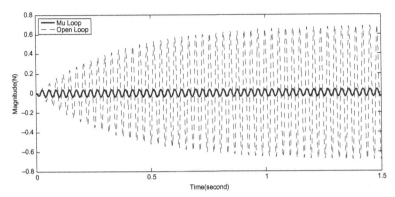

Figure 7.16 Open-loop and μ closed-loop response to a sinusoidal disturbance with a frequency of 33 Hz.

Figures 7.16 and 7.17 present the transient response to a harmonic disturbance input, and from the figures it can be seen that the μ controller or PI controller can effectively isolate the harmonic disturbance with a frequency of 33 Hz by more than 25.2 dB (94.5%).

For comparison, Figure 7.18 shows the open-loop response to a stochastic disturbance, which is normally distributed Gaussian white noise with mean zero and standard deviation 0.60. Simultaneously, the sensor noise is also considered, and the standard deviation of the sensor noise is 2% of the stochastic disturbance. Figures 7.19 and 7.20 shows the responses of the μ controller and the PI controller to stochastic disturbance and sensor noise. It can be seen that the standard deviations are attenuated by 11 dB (70%)

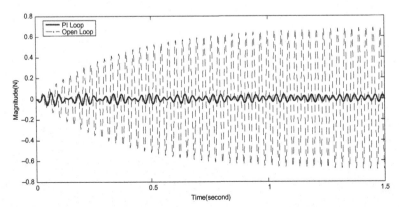

Figure 7.17 Open-loop and PI closed-loop response to a sinusoidal disturbance with a frequency of 33 Hz.

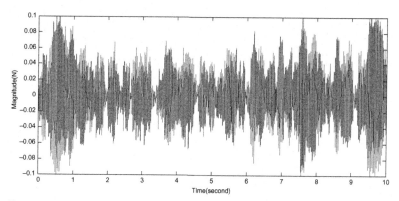

Figure 7.18 Open-loop response to stochastic disturbance.

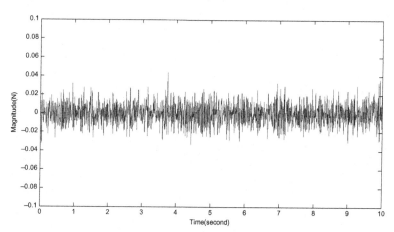

Figure 7.19 Closed-loop response of the μ controller to stochastic disturbance.

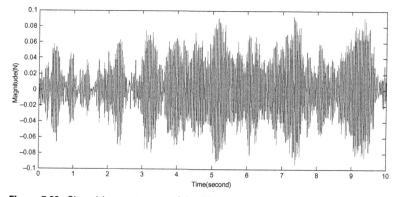

Figure 7.20 Closed-loop response of the PI controller to stochastic disturbance.

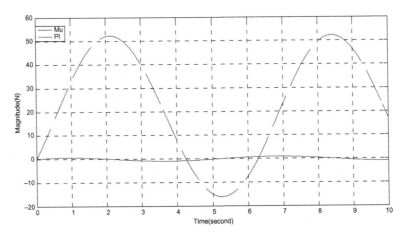

Figure 7.21 The magnitude for the PI and μ controllers with input r_0.

by the proposed μ controller, but the stochastic disturbance is amplified by 132% by the proposed PI controller.

The magnitudes for the PI and μ controllers are shown in Figure 7.21. The input r is a sinusoidal tracking force r_0. It can be seen that the magnitude for the PI controller is much larger than that for the μ controller. Additionally, with the PI controller, PZT actuators are more easily saturated because of the large response.

To verify the two requirements of the μ controller, another input signal is selected, and is made up of the tracking signal r_0, sinusoidal disturbance dist, stochastic disturbance, and the sensor noise. The open-loop response is shown on the left in Figure 7.22, from which it can be seen that the tracking signal is destroyed by the relatively small disturbance (5% of the tracking signal), but the

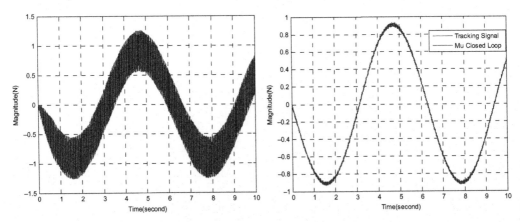

Figure 7.22 Open-loop and μ closed-loop response to the tracking signal with sinusoidal and stochastic disturbance.

closed-loop response gives a satisfactory result, as shown on the right in Figure 7.22.

7.5.2 Discussion

The robust stability margin for the h_∞ controller is 156%, less than the 629% for the μ controller, but its vibration isolation performance at resonance is better than that of the μ controller, and its order is lower too. The reduced controllers, by square root balanced model truncation, keep the robust stability compared with the original controllers. The time responses of the μ controller indicate that the harmonic disturbance with a frequency of 33 Hz is isolated by 25.2 dB (94.5%) using the proposed μ controller, the normally distributed disturbance with mean zero and standard deviation 0.2 is isolated by 11 dB (70%) using the proposed μ controller, a tracking command is kept with attenuation less than 0.2 dB, and harmonic and stochastic disturbance are rejected.

7.6 Conclusions

In this chapter we have presented multiobjective robust H_∞ and μ synthesis for active vibration control of the flexure struts of a Stewart platform, presented coupling analysis between six struts, and derived a dynamic model with the rigid mode removed for active vibration control. The corresponding robust H_∞ and μ synthesis controllers were given considering the noise of sensors and the coupling of the other five struts. The simulation results indicate that two requirements are achieved simultaneously with the proposed robust controller.

References

[1] H. Chen, E. Hospodar, B. Agrawal, Development of a hexapod laser-based metrology system for finer optical beam pointing control, AIAA International Communication Satelite Systems Conference and Exhibit, Paper no. 2004-3146, 2014.

[2] M. McMickell, T. Kreider, Optical payload isolation using theminiature vibration isolation system (MVIS-II), in: Proceedings of SPIE Conference on Industrial and Commercial Applications of Smart Structures Technologies 2007-652703, 2007.

[3] H.J. Chen, Payload pointing and active vibration isolation using hexapod platforms, 44th AIAA/ASME/ASCE/AHS Structures, Structural Dynamics, and Materials Conference, Norfolk, Virginia, USA, Paper no. 2003-1643, 2003.

[4] T. Hindle, T. Davis, J. Fischer, Isolation, pointing, and suppression (IPS) system for high performance spacecraft, in: Proceeding of SPIE Conference on Industrial and Commercial Applications of Smart Structures Technologies 2007-652705, 2007.

[5] V. Ford, Terrestrial planet finder coronagraph observatory summary, NASA Report Document ID 20060043653, 2005.

[6] M. Winthrop, R. Cobb, Survey of state of the art vibration isolation research and technology for space applications, in: Proceedings of SPIE Conference on Smart Structures and Materials 5052 (2003) 13-26.

[7] Z. Geng, S. Haynes, Six degree-of-freedom active vibration control using the stewart platforms, IEEE Trans. Control Syst. Technol. 2 (1) (1994) 45-53.

[8] E.H. Anderson, J.P. Fumo, R.S. Ervin, Satellite ultra-quiet isolation technology experiment (SUITE), in: Proceedings of IEEE Aerospace Conference, vol. 4, 2000, pp. 219-313.

[9] J.E. McInroy, Properties of orthogonal Stewart platforms, in: Proceedings of SPIE Conference on Smart Structures and Integrated Systems, 2003, pp. 591-602.

[10] Y.X. Chen, E. McInroy, Decoupled control of flexure jointed hexapod using estimated joint space mass inertia matrix, IEEE Trans. Control Syst. Technol. 12 (3) (2004) 413-421.

[11] J. Spanos, A soft 6-axis active vibration isolator, in: Proceedings of American Control Conference, 1995, pp. 412-416.

[12] D. Thayer, Multi sensor control for 6-axis active vibration, PhD thesis, University of Washington 1998.

[13] D. Thayer, C. Campbell, Six-axis vibration isolation system using soft actuators and multiple sensors, J. Spacecr. Rocket. 39 (2) (2002) 206-212.

[14] A. Preumont, A. Francois, Force feedback versus acceleration feedback in active vibration isolation, J. Sound Vib. 257 (4) (2002) 605-613.

[15] E. McInroy, Modeling and design of flexure jointed stewart platforms for control purposes, IEEE/ASME Trans. Mechatron. 7 (1) (2002) 95-99.

[16] R. Cobb, Vibration isolation and suppression system for precision payloads in space, Smart Mater. Struct. 8 (6) (1999) 798-812.

[17] S. Hauge, E. Campbell, Sensors and control of a space-based six-axis vibration isolation system, J. Sound Vib. 269 (4) (2004) 913-931.

[18] A. Joshi, System identification and multivariable control design for a satellite ultra-quiet isolation technology experiment (SUITE), MS thesis, Texas A and M University, 2002.

[19] A. Joshi, W. Kim, Modeling and multivariable control design methodologies for hexapod-based satellite vibration isolation, J. Dyn. Syst. Measur. Control 127 (4) (2005) 700-704.

[20] K. Gawronski, Advanced Structural Dynamics and Active Control of Structures, Springer, New York, 2004.

[21] S. Skogestad, I. Postlethwaite, Multivariable Feedback Control: Design and Analysis, second ed., John Wiley and Sons, Chichester, 2005.

[22] J. Doyle, K. Glover, P. Khargonekar, B. Francis, State-space solutions to standard H_2 and H_∞ control problems, IEEE Trans. Autom. Control 34 (8) (1989) 831-847.

[23] K.M. Zhou, J. Doyle, Essentials of Robust Control, Prentice Hall, Upper Saddle River, 1998.

[24] J. Doyle, Analysis of feedback systems with structured uncertainties, IEEE Proc. D Control Theory Appl. 129 (1982) 242-250.

[25] P.M. Yong, Robustness with parametric and dynamic uncertainty, PhD thesis, California Institute of Technology, 1993.

[26] D. Enns, Model reduction for control system design, PhD thesis, Stanford University, 1984.

[27] G. Obinta, Model Reduction for Control System Design, Springer, Berlin, 2000.

[28] D.W. Gu, P. Petkov, M. Konstantinov, Robust Control Design with MATLAB, Springer-Verlag, London, 2005.

8

CONCLUSIONS

Modeling and precision control of smart systems with hysteresis are widely demanded. In this book we have presented the fundamental modeling and control approaches for engineers. Furthermore, sufficient experimental demonstrations have been provided.

In Chapter 2, the fundamentals of smart systems with hysteresis were presented. Mechanical vibration and capacitor dynamics were illustrated. The Preisach model, the Preisach plane, and properties of the static hysteresis were investigated in detail. The behavior of the Preisach hysteresis and that of the linear dynamics phase delay were compared. Further, the closed-loop response of smart systems with hysteresis was presented. To model smart hysteresis at broadband frequencies, a composite representation was presented.

In Chapter 3, we mainly focused on smart actuators. The hysteresis modeling of smart hysteresis was investigated. Static hysteresis, composite hysteresis, and the persistent-excitation problem were presented. Simple identification approaches were also provided. Finally, the response components of smart actuators were illustrated.

In Chapter 4, the complete modeling and identification were developed. To begin with, the experimental setup of soft piezoelectric smart systems was presented. Then, the multifield hysteretic dynamics of the piezoelectric platform was derived. After the modeling, the comprehensive identification of the hysteretic dynamics was presented. The experimental results demonstrate the effectiveness of the proposed modeling and identification approaches. The complete modeling method for the multifield dynamics in soft smart systems was extended to hard piezoelectric smart systems.

Modeling and Precision Control of Systems with Hysteresis
http://dx.doi.org/10.1016/B978-0-12-803528-3.00008-2

In Chapter 5, the multirate composite controller with limited digital signal processors was presented. The fast-sampling feedforward control was built with use of the identified hysteretic dynamics. The slow-sampling discrete H_∞ controller was designed with use of the nonhysteretic dynamics. The proposed composite controller was implemented in a piezoelectric stage with a digital signal processor platform. Fast and accurate tracking performance was verified in experiments.

In Chapter 6, a case study of a piezoelectric platform system was presented. The measurement strategy was investigated to provide effective measurement with high bandwidth and accuracy. A comprehensive design of the composite control strategy was developed. The effectiveness of precision angle motion was validated in experiments.

In Chapter 7, we presented multiobjective robust control and synthesis for active vibration control of the flexure struts of a Stewart platform. The dynamic model was derived with the rigid mode removed for active vibration control. The robust controller and the μ synthesis controller were given considering the sensor noise and couplings. The simulation results indicate that the requirements are achieved simultaneously with the proposed robust controller.

The proposed modeling and precision control approaches in this book are beneficial for further developments in engineering. For instance, to achieve high-accuracy tracking and positioning, model-based modern controllers can be designed with use of the hysteretic dynamics. Especially, modeling and control suggestions have been provided for engineers in this book.

APPENDIX A

A.1: Persistent-excitation Condition Under Designed Input Signals and Sampling Rules

A.1.1: Proposed Inputs and Sampling Law

The section illustrates that the persistent-excitation (PE) condition is satisfied by use of the input in (A.1):

$$u(t) = \frac{P(t)}{2}(1 - \cos(\omega_r t)), \qquad (A.1)$$

$$P(t) = \text{fix}\left(1 + \frac{t - \text{fix}(t/T)T}{T_{\text{sub}}}\right) \cdot \delta, \qquad (A.2)$$

where fix(x) is a function to round x to its nearest integer toward zero and δ is given in (A.3),

$$\delta = u_{\max}/L, \qquad (A.3)$$

where T is the whole period, L is the discretization level, and $u_{\max}$ is the maximum voltage in the identification.

The sampling is given in (A.4):

$$\begin{cases} t_{i,j} = \frac{1}{\omega_r}\arccos(1 - \frac{2j\delta}{P(t)}) + t_{i,1}, \\ t_{i,2i-j} = \frac{1}{\omega_r}[2\pi - \arccos(1 - \frac{2j\delta}{P(t)})] + t_{i,1}, \end{cases} \qquad (A.4)$$

where $j = 1, 2, \cdots, P_i/\delta_s$.

For instance, let $N = 1$, $\delta = 1$, $\delta_s = 1$, $\omega_r = \pi$, and $T = 8$. The density values of 10 grids need to be identified. According to the input signal and the sampling rule, partial sampling points are collected and used. Figure A.1 illustrates the sampling points corresponding to the discretization points on the Preisach plane in the first four periods. It can be seen that 10 samples are collected in the first four periods.

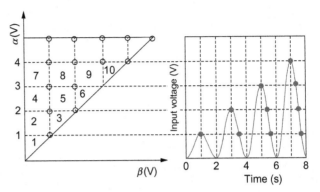

Figure A.1 Illustration of sampling points.

According to the discretization of the Preisach model, matrix A for the 10 sampling points is obtained as shown in (4.20). Furthermore, one nonsingular matrix E can be used to premultiply matrix A to exchange the columns of A, such that the following EA is achieved:

$$
A = \begin{bmatrix}
1 & 0 & 0 & 0 & 0 & 0 & 0 & 0 & 0 & 0 \\
1 & 1 & 1 & 0 & 0 & 0 & 0 & 0 & 0 & 0 \\
1 & 1 & 0 & 0 & 0 & 0 & 0 & 0 & 0 & 0 \\
1 & 1 & 1 & 1 & 1 & 1 & 0 & 0 & 0 & 0 \\
1 & 1 & 1 & 1 & 1 & 0 & 0 & 0 & 0 & 0 \\
1 & 1 & 0 & 1 & 0 & 0 & 0 & 0 & 0 & 0 \\
1 & 1 & 1 & 1 & 1 & 1 & 1 & 1 & 1 & 1 \\
1 & 1 & 1 & 1 & 1 & 1 & 1 & 1 & 1 & 0 \\
1 & 1 & 1 & 1 & 1 & 0 & 1 & 1 & 0 & 0 \\
1 & 1 & 0 & 1 & 0 & 0 & 1 & 0 & 0 & 0
\end{bmatrix}, \quad
EA = \begin{bmatrix}
1 & 0 & 0 & 0 & 0 & 0 & 0 & 0 & 0 & 0 \\
1 & 1 & 0 & 0 & 0 & 0 & 0 & 0 & 0 & 0 \\
1 & 1 & 1 & 0 & 0 & 0 & 0 & 0 & 0 & 0 \\
1 & 1 & 0 & 1 & 0 & 0 & 0 & 0 & 0 & 0 \\
1 & 1 & 1 & 1 & 1 & 0 & 0 & 0 & 0 & 0 \\
1 & 1 & 1 & 1 & 1 & 1 & 0 & 0 & 0 & 0 \\
1 & 1 & 0 & 1 & 0 & 0 & 1 & 0 & 0 & 0 \\
1 & 1 & 1 & 1 & 1 & 0 & 1 & 1 & 0 & 0 \\
1 & 1 & 1 & 1 & 1 & 1 & 1 & 1 & 1 & 0 \\
1 & 1 & 1 & 1 & 1 & 1 & 1 & 1 & 1 & 1
\end{bmatrix}
$$

(A.5)

Matrix EA is lower triangular and full rank. Thus, matrix A is also nonsingular and full rank because of the nonsingularity of E. According to the sampling law, the special sampling points are reserved in matrix A. If the detailed disretization is applied, matrix $A^{T}A$ is still nonsingular and full rank. Finally, the PE condition in the estimation equation is satisfied.

A.1.2: Persistent-Excitation Condition with Standard Inputs and Sampling

Next, the input signals with constant amplitude and the time-based sampling (uniformly over time) are used for comparison.

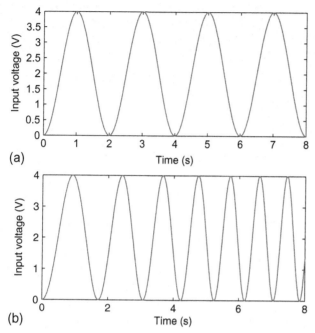

Figure A.2 Two standard inputs. (a) The input signal with a fixed amplitude. (b) The input signal with varying frequencies.

In this section, two inputs are considered. Figure A.2(a) shows the input with a constant frequency of 0.5 Hz and a constant amplitude of 4 V. Figure A.2(b) shows the input with a constant amplitude of 4 V but varying frequencies. The sampling is based on time and the sampling interval is set to 0.01 s. According to the Preisach discretization in Section A.1.1, matrix A is obtained. In both cases, 800 points are collected and used in 8 s. However, the rank of $A^T A$ in both cases is 7. The PE condition in both cases is not satisfied.

A.2: Inversion of the Preisach Hysteresis Model

The inversion of the identified Preisach hysteresis model $\hat{\Gamma}^{-1}$ is shown in Algorithm 1. δ_s is the iteration size, q is the maximum iteration number, v_r is the reference trajectory, which is also the output of $\hat{G}^{-1}$, and u_{ff} is the feedforward control signal. In the piezoelectric stage experiment, δ_s and q are set to 0.2 and 10, respectively.

ALGORITHM 1 INVERSION OF THE IDENTIFIED PREISACH HYSTERESIS MODEL

if $v_{\mathrm{r}}(k) < \hat{v}(k)$ **then**

    $u_{\mathrm{ff}}(k+1) = u_{\mathrm{ff}}(k) - i\delta_{\mathrm{s}}, \qquad i = 1, \cdots, q$

    $\hat{v}(k+1) = \hat{\Gamma}(u_{\mathrm{ff}}(k+1))$

    if $|v_{\mathrm{r}}(k) - \hat{v}(k+1)| \leq \delta_{\mathrm{s}}$ **then**

      break

    end if

else

    if $v_{\mathrm{r}}(k) > \hat{v}(k)$ **then**

      $u_{\mathrm{ff}}(k+1) = u_{\mathrm{ff}}(k) + i\delta_{\mathrm{s}}, \qquad i = 1, \cdots, q$

      $\hat{v}(k+1) = \hat{\Gamma}(u_{\mathrm{ff}}(k+1))$

      if $|v_{\mathrm{r}}(k) - \hat{v}(k+1)| \leq \delta_s$ **then**

        break

      end if

    else

      $u_{\mathrm{ff}}(k+1) = u_{\mathrm{ff}}(k)$

    end if

end if

INDEX

Note: Page numbers followed by f indicate figures and t indicate tables.

Printed in the United States
By Bookmasters